D1312009

Marian Anton is Associate Professor at Central Connecticut State University and a recipient of the *Gheorghe Lazar Prize of the Romanian Academy*. He started his teaching career as a middle school teacher.

❧

Karen Santoro is Assistant Professor, the Developmental Mathematics Coordinator, and the Bridges Program Director at Central Connecticut State University. She co-authored the 2nd edition of *Math Connections, A Secondary Mathematics Curriculum*, funded by the NSF.

Introduction

This book has been designed for a one semester course for students entering college at the developmental level. Its purpose is to enhance students' algebraic, critical reasoning, and communication skills, and to promote a growth mindset towards learning. In order to achieve our goals, we envision a guided inquiry approach through collaborative exploration, concept formation, and application where students learn the algebraic concepts and their meaning in real-world contexts. We focus on fewer concepts, mainly linear, exponential, and quadratic functions, but treat them conceptually, in more depth, and interconnected for better retention.

The curriculum and pedagogy for the course were first developed and tested in class through slides, handouts, and online quizzes. However, the lack of a textbook proved to be an unnecessary burden for both the instructors and the students. Unlike lecture notes, we decided to treat it as a true reference book, including all technical definitions and formulas in a rigorous manner. As such, there are details in the book that are more technical than is appropriate for our intended student audience, and comments to this effect will be included in the book's teaching notes.

The movie Groundhog Day is a good metaphor for our book's concept. Each day the lead character wakes up on the same day, Groundhog Day, repeating the same routine while struggling with a situation. He starts to predict what is going to happen each day, he learns a little bit more, and gets closer to solving the problem. By the end he understands what he needs to do. Our students will be taken through a similar experience. Each section starts with a launch exploration based on the common threads of the same routine: when given a situation, they identify the variables, make tables, draw graphs, search for patterns, and write equations. Students may struggle to identify variables in problems and situations early in the course, but they are asked to do so in EVERY problem in EVERY unit so they can master that skill by the end.

The role of developmental mathematics is to prepare students to succeed in college-level courses by improving their quantitative literacy skills as defined by the Association of American College and Universities as "a 'habit of mind,' competency, and comfort in working with numerical data ... ability to reason and solve quantitative problems from a wide array of ... everyday life situations." Curriculum Renewal Across the First Two Years subcommittee of the Mathematical Association of America recommends emphasis on conceptual understanding, problem solving skills, mathematical modeling, and communication, and this book implements those recommendations in an introductory-level algebra class for college students.

Acknowledgements

This book could not have been written without the feedback from both nominal and anonymous instructors at CCSU across disciplines. In alphabetical order here is the nominal list of people who answered our surveys: Adam Allan, James Arena, Mihai Bailesteanu, Roger Bilisoly, Alicia Bray, Tom Burkholder, Nelson Castaneda, Marianne Chamberlain, Penny Coe, Richard Coe, Philip Crockett, Igor Ferdman, Cheryl Fox, Shelly Jones, Robin Kalder, Wojciech Kolc, Bujar Konjusha, Fred Latour, Sally A. Lesik, Eran Makover, Dan Miller, Oscar Perdomo, Stewart Rosenberg, Maria Roxas, Elizabeth Salgado, Pinchas Schreiber, Thomas Tiezzi, Tony Weisgram, Barry Westcott, and Arben Zeqiraj.

Our special thanks go to Lubna Abdulrahman, Nancy Barger, Cheryl Fox, Judy Hodgson, Gil Linder, Robin Kalder, Wojciech Kolc, Judy Marzi, and Ann Marie Spinelli for reading drafts of the manuscript and providing invaluable input. We also thank our external reviewers for their time, effort, and timely response.

A Suggested Timeline (15 Weeks)

Section/Content	# of classes
Unit 0 - Review Concepts	
0.1 Order of Operations	0-1
0.2 Perimeter, Area, and Volume	1-2

perimeter & area of basic shapes and their composites; volume of prisms & cylinders; Pythagorean Theorem

Unit 1 - Variables and Functions	
1.1 Variable versus Constant Quantities	1

identify quantities, including rates, choose units of measure; distinguish between constant & variable quantities; choose 'smart' labels; classify variables as continuous or discrete

1.2 Translating English to Algebra and Back	1-2

translate English into symbolic algebra and vice versa using expressions, equations & inequalities; perform quantitative analysis of a situation including units & conversions; write and create formulas using geometry or general knowledge that are relevant to the problem

1.3 Algebraic Problem Solving with Variables	2 -3

set up systems of linear & power equations in two or three variables in contexts; solve them by substitution; use square and cubic roots; check solutions & verify inequality constraints; problems may involve inconsistent, extraneous/invalid or multiple solutions

1.4 Relationships between Variables - Introduction to Functions	1

define functions as sets of ordered pairs; give a finite domain/range using set notation; use function notation $y = f(x)$ in context; switch between ordered pairs and function notation; represent functions by tables & scatterplots; distinguish them from relations

1.5 Analyzing Real World Functions (Models)	2

represent a function as a set of ordered pair solutions to an equation; find/interpret 'x'-intercepts, 'y'-intercept/initial value; give a real-world domain & range; use interval notation; choose scales; identify increasing/decreasing functions; use a function graph/equation to answer questions about a situation it models

Test 1 (Unit 0 & Unit 1)	
Unit 2 - Linear Models	
2.1 Introduction to Linear Relationships	1

recognize linear functions from tables by constant first order differences; define linear functions as functions with a constant rate of change between any two points; find this rate from tables; interpret the sign of the constant rate; write a linear equation $y = mx + b$ in context

2.2 Analyzing Linear Functions ($y = mx + b$)	1-2

define the slope of a line; use slope triangles; interpret the slope of a linear graph as the constant rate of change; write a linear equation given two data points; interpret slope as direction and steepness; constant function; discuss vertical, horizontal, and parallel lines

2.3 Creating & Using Linear Models ($y = mx + b$)	1-2

create linear models by writing equations from two data points; complete a linear graph; use linear graphs/equations to answer questions about situations (interpolations and extrapolations); analyze assumptions and limitations of linear models (domain and range)

Unit 3 - Exponential Models	
3.1 A New Pattern - Introduction to Exponential Functions	1

recognize exponential functions from tables by constant multipliers/ratios; define exponential functions by equations of the form $y = a(b)^x$; distinguish between exponential, linear, & power; write an exponential function equation in context or given two consecutive data points

3.2 Features of Exponential Function Graphs	1-2

define exponential growth/decay functions; find the 'y'-intercept (initial value); discuss the asymptotic behavior; give a real world domain/range in context; choose scales using scientific notation; complete an exponential graph; compare linear and exponential models

3.3 Problem Solving with Exponentials	2

define the relative growth/decay rate; find a growth/decay factor given a growth/decay rate and vice versa; create exponential models in context; define the logarithm $x = \log_b y$ and use exponential function equations to answer questions (interpolations and extrapolations)

Test 2 (Unit 2 & Unit 3)	
Unit 4 - Quadratic Models	
4.1 Another New Pattern - Introduction to Quadratic Functions	1

connect algebraic to geometric concepts of 'squaring' a number; recognize quadratic functions by constant second order differences; define quadratic functions by equations of the form $y = ax^2 + bx + c$ and recognize the shape of their graphs as parabolas; briefly discuss properties of parabolas and their applications

4.2 Algebraically Finding Important Features of Parabolas	2

identify the important features of a parabola: direction, vertex - maximum or minimum point, 'y'-intercept, 'x'-intercept(s), axis of symmetry; use vertex formula $x = -b/2a$; discuss mirror points; use the Quadratic Formula; solve simple quadratic equations by factoring (lightly); discuss the number of real solutions

4.3 Analyzing Quadratic Functions   2

use quadratic function equations to answer questions about revenue, profit and cost in business, projectile motion in physics; give a real world domain/range of a quadratic model; interpret the important features of a quadratic function in a given context

Test 3 (Unit 4 cummulative)

Cummulative Final Exam

A Survey for the First Day of Classes

While reading this book think about the statements below and discuss with peers and teachers how your views affect your learning and success in mathematics.

- I like math.

- Math is facts and procedures.

- I am quick to understand math.

- I enjoy being math challenged.

- Math answers are right or wrong.

- Math is boring.

- Math people answer quickly.

- Math is confusing.

- It's important to think about math ideas.

- I look forward to my math lessons.

- It's important to memorize math.

- I can tell if math answers make sense.

- A bad grade means I am not smart.

- I believe that I can do well in math.

- When I face difficult math I give up.

- When I make a mistake I feel bad.

- Math is creative.

- It is important in math to be fast.

- There are limits in math abilities.

- With effort I can succeed in math.

- Math makes me feel afraid.

- You cannot change math intelligence.

- Math ideas have lots of connections.

- It is helpful to talk about math.

- It's only one way to solve a problem.

- I like to solve complex problems.

Contents

Introduction   **v**

0 Review Concepts   **1**
   0.1   Order of Operations 1
   0.2   Perimeter, Area, and Volume 6

1 Variables and Functions   **15**
   1.1   Variable versus Constant Quantities 15
   1.2   Translate English to Algebra and Back 22
   1.3   Algebraic Problem Solving with Variables 32
   1.4   Relationships between Variables - Introduction to Functions . . 46
   1.5   Analyzing Real World Functions (Models) 56

2 Linear Models   **67**
   2.1   Introduction to Linear Relationships 67
   2.2   Analyzing Linear Functions ($y = mx + b$) 77
   2.3   Creating and Using Linear Models ($y = mx + b$) 90

3 Exponential Models   **101**
   3.1   A New Pattern - Introduction to Exponential Functions 101
   3.2   Features of Exponential Function Graphs 111
   3.3   Problem Solving with Exponentials 121

4 Quadratic Models   **133**
   4.1   Another New Pattern - Introduction to Quadratic Functions . . . 133
   4.2   Algebraically Finding Important Features of Parabolas 141
   4.3   Analyzing Quadratic Functions 151

Index   **159**

Unit 0

Review Concepts

0.1 Order of Operations

Launch Exploration

Evaluate the following expressions without using a calculator or reading ahead:

(0.1) $$48 \div 2(3+1), \qquad -6^2 \div 2 * 3 + 4, \qquad 9 - 3 \div \frac{1}{3} + 1.$$

Then compare your answers with some classmates. If you get different values try to understand why. What questions arise? Write them down.

Key Questions

Which operation should you do first?

Which one comes next?

Which operation should you perform last?

Table 0.1

Order of Operations - A Case Study

Suppose that you have to evaluate the following expression:

(0.2) $$\frac{-3^2 - \sqrt{(-8)^2 - 5 \cdot (-3) \cdot (-1)}}{2 \cdot (-2)}.$$

When you find more than one operation in an expression like this, you should ask the questions on the margin in **Table 0.1. The Order of Operations** answers these questions! But first, what are the operations on numbers? In no particular order, the **operations** are *raising to powers, extracting roots, addition, subtraction, multiplication, division, taking the opposite,* etc.

Notice that *parentheses are not operations!* You may already know that the Order of Operations should be done according to the **PEMDAS** rule, often remembered by "**P**lease **E**xcuse **M**y **D**ear **A**unt **S**ally", but that doesn't help if you don't completely understand it. If you already know what each letter abbreviates, that's a great start but *it's not the full story.*

PEMDAS - NOT the full story

P - Parentheses

E - Exponents

M - Multiplication

D - Division

A - Addition

S - Subtraction

Order of Operations - The Full Story

P(arentheses) represent *all grouping symbols.* **The first step** in the rule means to **execute the operations inside grouping symbols.** If you have grouping symbols

1

nested one inside another, **work from the inside out**. Note that some grouping symbols are also operations and others are not. You can think of many grouping symbols as having parentheses *'built in'*. The parentheses inside these symbols are implied. We will write them here in **purple** to make it more clear how these symbols group things:

Fact 0.1 *Grouping Symbols - Implied Parentheses*

( ), [( )]	–	Parentheses and brackets are pure grouping symbols, not operations
|( )|	–	Absolute value symbol
$\sqrt{(\ )}$, $\sqrt[3]{(\ )}$, ...	–	Radical or root symbols
$\dfrac{(\ )}{(\ )}$	–	Division bar symbol

☞ Roots are actually exponents $\sqrt[3]{2} = 2^{1/3}$ and division is actually multiplication $5 \div 7 = 5 \cdot 7^{-1}$ or $5/7 = 5 \cdot (1/7)$.

E(xponents) represent *exponents and roots* since these are really on the same level in the Order of Operations. **The second step** in the rule means to **raise all bases to the powers indicated by their exponents and extract all the roots.** If you have more than one of these operations in an expression, do them in order **from left to right**. Remember that exponents and roots are inverses of each other and so they are on the same level in the Order of Operations.

M(ultiplication) and **D**(ivision) are really on the same level in the Order of Operations. **The third step** in the rule means to **perform mutiplications and divisions.** Also taking the opposite of an expression by changing to the opposite sign is the same as multiplication by (-1) so it's on this level. Let's call this operation **"negation"** and treat it like multiplication. For example, the results of negation could be negative or positive numbers:

☞ Once we have established that negation is like multiplication, in practice we don't mutiply, but we write $-5^2 = -(5^2) = -(25) = -25$.

Notice that $-5^2 = -25$ is not equal $(-5)^2$ since

$(-5)^2 = (-5)(-5) = 25$.

(0.3)
$$-5^2 = (-1)*(5^2) = (-1)*(25) = -25,$$
$$-(3-10) = (-1)*(3-10) = (-1)*(-7) = 7.$$

If you have more than one of these operations in an expression, do them in order **from left to right**. Remember that multiplication and division are inverses of each other and so they are on the same level in the Order of Operations.

A(ddition) and **S**(ubtraction) are really on the same level in the Order of Operations. **The last step** in the rule means to **perform additions and subtractions.** If you have more than one of these operations in an expression, do them in order **from left to right**. Remember that addition and subtraction are opposites of each other and so they are on the same level in the Order of Operations. In conclusion, the full story can be summarized in the table below.

☞ Subtraction is an algebraic addition $4 - 6 = 4 + (-6)$.

> **Fact 0.2** *GERMDAS*
>
> | | | | |
> |---|---|---|---|
> | **First:** | **G** | Perform operations inside **Grouping Symbols** | Inside to Outside |
> | **Second:** | **E/R** | Perform **Exponents and Roots** | Left to Right |
> | **Third:** | **M/D** | Perform **Multiplications and Divisions** | Left to Right |
> | **Last:** | **A/S** | Perform **Additions and Subtractions** | Left to Right. |

Example 0.1

Evaluate the expression (**0.2**) from the previous subsection:

$$\frac{-3^2 - \sqrt{(-8)^2 - 5 \cdot (-3) \cdot (-1)}}{2 \cdot (-2)}.$$

Solution: We'll use the Order of Operations, of course! Start by identifying the *grouping symbols!* Recall that the radical and division bar symbols are grouping symbols. *Show the grouping by inserting your own parentheses.* Using **green** parentheses for the radical grouping and **purple** parentheses for the division bar grouping, we can put the two together:

(**0.4**)
$$\frac{\left(-3^2 - \sqrt{((-8)^2 - 5 \cdot (-3) \cdot (-1))}\right)}{(2 \cdot (-2))}. \qquad \text{(Nested Grouping)}$$

Since we have nested groupings, we simplify them from the inside out. Observe that within parentheses such as (-8), (-3), etc., there are no operations to be performed. The parentheses only contain the negative numbers -8, -3, etc. Since there is nothing to simplify inside these parentheses, we simplify inside the **green** parentheses first. Grouped under the radical sign we have 1 exponent, 2 multiplications, and 1 subtraction. Do them in order according to **GERMDAS**. If there are more than one at the same level, do them from left to right.

$$
\begin{aligned}
((-8)^2 - 5 \cdot (-3) \cdot (-1)) &= (\mathbf{64} - 5 \cdot (-3) \cdot (-1)) && \text{(Exponent)} \\
&= (64 - (\mathbf{-15}) \cdot (-1)) && \text{(1st Multiplication)} \\
&= (64 - \mathbf{15}) && \text{(2nd Multiplication)} \\
&= \mathbf{49}. && \text{(Subtraction)}
\end{aligned}
$$

Now the expression (**0.4**) is still grouped by the division bar. Using the result above, we can *choose* to work in the numerator next and follow the same procedure:

$$
\begin{aligned}
\left(-3^2 - \sqrt{((-8)^2 - 5 \cdot (-3) \cdot (-1))}\right) &= \left(-3^2 - \sqrt{\mathbf{49}}\right) && \text{(Substitution)} \\
&= (-(\mathbf{9}) - \mathbf{7}) && \text{(Exponent \& Root)} \\
&= (\mathbf{-9} - 7) && \text{(Negation)} \\
&= \mathbf{-16}. && \text{(Subtraction)}
\end{aligned}
$$

(**0.5**)
$$\left(-3^2 - \sqrt{((-8)^2 - 5 \cdot (-3) \cdot (-1))}\right) = -16. \qquad \text{(Numerator)}$$

(Denominator)

Finally, we have one multiplication left in the denominator of (**0.4**):

(**0.6**) $(2 \cdot (-2)) = -4.$

By substituting (**0.5**) and (**0.6**) in (**0.4**), the value of the initial expression (**0.2**) is

$$\frac{-3^2 - \sqrt{(-8)^2 - 5 \cdot (-3) \cdot (-1)}}{2 \cdot (-2)} = \frac{-16}{-4} = 4.$$

(Division)

☞ PEMDAS or GERMDAS can
be thought of as machine
language acronyms for the
Order of Operations used by
calculators.

The Order of Operations is one of the most important things to know and to *understand* in algebra! But it's just as important to recognize when to use it. We need the Order of Operations to simplify algebraic expressions, solve algebraic equations, and do numerical evaluations with a **calculator**.

How to use a graphing calculator

To evaluate the expression (**0.2**) you push the following keys:

☞ There are many ways to
break down long sequences
into smaller parts.

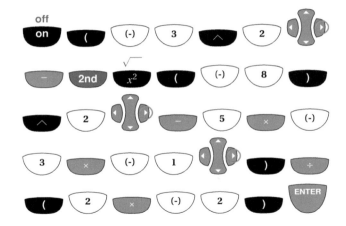

and you get the result 4. Identify the keys above, if any, which correspond to the parentheses *inserted* in the expression (**0.4**).

GERMDAS

G - Grouping Symbols

E/R - Exponents/Roots

M /D- Multiplication/Division

A/S - Addition/Subtraction

Wrapping Up

GERMDAS is just a memory aid for the Order of Operations. Parentheses alone are not operations. Exponents and roots are on the same level as are multiplication and division as well as addition and subtraction. Negation means taking the opposite (sign) and is on the same level as multiplication. Same level operations are performed from LEFT to RIGHT and nested groupings are performed from the inside out. If groupings are not nested you can choose which one to work out first.

So when you are doing algebra and ask "What should I do next?", think about the Order of Operations to find the answer! You have now learned the *full story* about the Order of Operations. It is time to practice!

Exercises

Exercises for 0.1 Order of Operations

P0.1 Can the evaluation of a well-defined expression lead to multiple answers? What is the role of **GERMDAS**?

P0.2 Evaluate the expressions (**0.1**) and compare the results with your initial answers. Should you apply **PEMDAS** *strictly* by doing multiplication before division and addition before subtraction?

P0.3 Evaluate the following expressions without a calculator.

(a) $\dfrac{\sqrt{25}}{2^2+1}$

(b) $-5^2-(3+4)^2$

(c) $-5(2-5)^2+6(-1)^2$

(d) $\sqrt{(6-2)^2}+\dfrac{8-2}{-6+3}$

(e) $\dfrac{(-7-2)^2}{7}$

(f) $5(2)^3$

(g) $-3^2+(-2)^4$

(h) $-(3-5)^3+2\cdot3^2$

(i) $(-21+4\cdot5)^2$

(j) $-5^2+3(-7)-1$

(k) $3^2-4(2)(-2)$

(l) $-(-7+3)-8\div2^2$

(m) $-7^2-4\cdot1^2$

(n) $\dfrac{12-\sqrt{3^2+5(-1)}}{4}$

(o) $\dfrac{-(1+3)^2+4}{2}$

(p) $\dfrac{-(-1)+\sqrt{(-1)^2-4(2)(-3)}}{2(2)}$

(q) $-8^2+(-3)^3$

(r) $\dfrac{8+6(-2+1)}{2}$

(s) $6\div2(1+2)$

P0.4 What is the outcome difference between
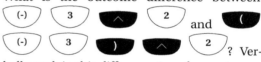
and
? Verbally explain this difference to a classmate.

P0.5 Check your answers in problem P0.3 by using a graphing calculator.

0.2   Perimeter, Area, and Volume

Launch Exploration

What is the perimeter of a shape? What is the area of a shape? What is the difference between the two? Explain it in your own words. Work together with peers and do the following tasks. Identify the shape in **Fig 0.1** and construct:

1. A rectangle having the same area as the given shape. How many such rectangles are there? How many of them are squares?

2. A rectangle having the same perimeter as the given shape. How many such rectangles are there? How many of them are squares?

Discuss the units of measure for the perimeter and area. Is it true that two shapes with the same area must have the same perimeter? Is the converse true?

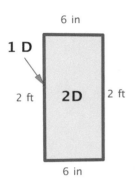

Figure 0.1 Perimeter/Area.

The Perimeter of a 2D Shape

A *distance* or *length* is measured in units such as inches, yards, feet, miles, centimeters, meters, kilometers, etc.

> **Definition 0.3 *Perimeter***
>
> The **perimeter** of a two dimensional shape is the distance traced along its one-dimensional boundary.

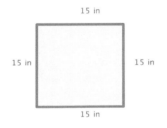

Figure 0.2 Perimeter = 60 in.

The perimeter is measured in units of length or distance. If the shape is bounded by various line-segments or circular arcs, the perimeter is the **sum** of those side- or arc-lengths expressed using a *common* unit of measure.

For example, a *rectangle* is a shape formed by four sides that make four interior right angles. The opposite sides of a rectangle have the same distance, the longer being called *length* and the shorter being called *width*.

(0.7)   **Perimeter of a rectangle = (length) + (width) + (length) + (width).**

In our case, the perimeter of the rectangle in **Fig. 0.1** can be calculated in inches by converting 2 feet to $2 \times 12 = 24$ inches:

Perimeter of the rectangle = (24 in) + (6 in) + (24 in) + (6 in) = 60 in.

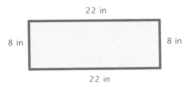

Figure 0.3 Perimeter = 60 in.

Examples of rectangles having the same perimeter of 60 inches are those of dimensions 3×27, 8×22, 12.7×17.3, or $c \times (30 - c)$ for any real number c between 0 and 30. A *square* is a rectangle with the four sides of equal length. The *only* square having the perimeter of 60 in must have its side length of $60 \div 4 = 15$ in.

$$\begin{aligned}
\text{Perimeter of the rectangle} &= (27\text{ in }) + (3\text{ in }) + (27\text{ in }) + (3\text{ in }) \\
&= (22\text{ in }) + (8\text{ in }) + (22\text{ in }) + (8\text{ in }) \\
&= (17.3\text{ in }) + (12.7\text{ in }) + (17.3\text{ in }) + (12.7\text{ in }) \\
&= (15\text{ in }) + (15\text{ in }) + (15) + (15\text{ in }) \\
&= (30 - c) + (c) + (30 - c) + (c) = 60\text{ in.}
\end{aligned}$$

We conclude that there are infinitely many rectangles with the same perimeter and only one of them is a square!

☞ By rigid motion we get many copies of the square, but we count only one.

The Area of a 2D Shape

The area of a shape is a two dimensional measure of the shape which is built up by covering the shape with unit squares and counting them. A **unit square** is a square of side 1 *unit* and its area is defined as 1 *unit*^{2}. So the units of measure for area are sq. meters, sq. inches, sq. yards, sq. ft, sq. miles, sq. cm, etc.

> **Definition 0.4** *Area*
>
> The **area** of a two dimensional shape is the *exact* number of unit squares that fit inside that shape. An area could be any positive real number.

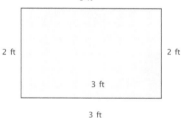

For example, to count the number of unit squares that fit inside a rectangle of dimensions 2 ft by 3 ft, we make a grid of unit squares of area 1 sq. foot each. Since the side of each such unit square is 1 foot long, we can fit 3 unit squares along the length and 2 rows of unit squares along the width of the rectangle. See **Fig 0.4**. By counting the unit squares we get the area of the rectangle as

$$\text{Area of the rectangle } = 3\text{ feet } \times 2\text{ feet } = 6\text{ sq. feet.}$$

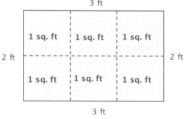

Figure 0.4 Area by counting unit squares.

We conclude that the area of a rectangle is the **product** of its length and width, both distances being expressed in a *common* unit of measure while the area is expressed using the square of that unit:

(0.8)   **Area of a rectangle = (length) × (width).**

In our case, the area of the rectangle in **Fig. 0.1** can be calculated in square feet by converting 6 inches to $6 \div 12 = 1/2$ feet and taking the product:

☞ **Area of a square**
   = (side length)^2

$$\text{Area of the rectangle} = (2\text{ feet }) \times \left(\tfrac{1}{2}\text{ feet }\right) = 1\text{ sq. foot.}$$

The *only* square having this area is the unit square with side length of 1 foot. We can show that the rectangle and the unit square have the same area by cutting the rectangle in two halves and pasting the two halves to a full square. See **Fig 0.5**. Examples of rectangles having the same area of 1 sq. ft are those of dimensions $3 \times (1/3)$, 1.25×0.8, or $c \times (1/c)$ for any positive real number c.

$$\text{Area of the rectangle } = (3)(1/3) = (1.25) \times (0.8) = c \times (1/c) = 1\text{sq. ft}$$

Perimeter vs. Area

dim (in)	perim (in)	area (in^2)
24 × 6	60	144
27 × 3	60	81
22 × 8	60	176
17.3 × 12.7	60	219.71
15 × 15	60	225
dim (ft)	perim (ft)	area (ft^2)
2 × (1/2)	5	1
3 × (1/3)	$6\frac{2}{3}$	1
1.25 × 0.8	4.1	1

Table 0.2

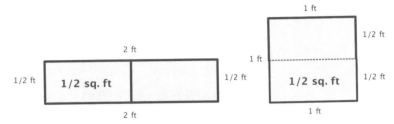

Figure 0.5 A rectangle having an area of 1 sq. ft.

Hence, we have infinitely many rectangles and only one square of area 1 sq. ft. As we see from the comparison **Table 0.2**, rectangles with the same perimeter may have different areas and vice versa. By the unit conversions 1 ft^2 = (12)^2 in^2 = 144 in^2 and 5 ft = 5 × 12 in = 60 in, each series starts with the same rectangle.

Perimeter and Area of a Triangle - Pythagorean Theorem

A *triangle* is a shape formed by three sides and three interior angles. By choosing one side-length of a triangle to be the *base* and the distance from the opposite vertex to the base to be the *height*, we have

(0.9)   **Area of a triangle $= \frac{1}{2}$ (height) × (base).**

To check this formula, we complete the triangle to a rectangle or a parallelogram having area twice as much as the area of the triangle. A *parallelogram* is a shape formed by four sides that are pairwise parallel and its area is given by height × base. See **Fig 0.6**.

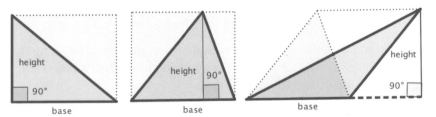

Figure 0.6 Completing a triangle.

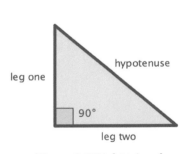

Figure 0.7 Right triangle

The longest side in a right triangle is called the *hypotenuse* and the other two sides are called *legs* such that their lengths satisfy the **Pythagorean Theorem**:

(0.10)   **(hypotenuse)^2 = (leg one)^2 + (leg two)^2.**

This equation is used to calculate side-lengths and heights of shapes on a grid.

Example 0.2 Find the area and perimeter of each shape in **Fig 0.8**; each square on the gird in part (a) represents 1 sq. mile and the shapes in parts (b) and (c) are shaded.

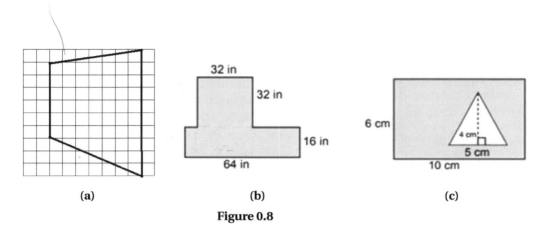

(a)   **(b)**   **(c)**

Figure 0.8

a) The shape can be decomposed into a 6×7 mid rectangle, a top right triangle with legs 1 and 7, and a bottom right triangle with legs 3 and 7 miles.

$$\text{Area } = (6 \times 7) + \tfrac{1}{2}(1 \times 7) + \tfrac{1}{2}(3 \times 7) = \left(6 + \tfrac{1}{2} + \tfrac{3}{2}\right) \times 7 = 8 \times 7 = 56 \text{ sq. miles}$$

Alternatively, notice that the shape is a trapezoid with bases 6, 10 and height 7 miles. The calculation above shows that the area of the trapezoid is the average $(6 + 10)/2 = 8$ miles of the bases times its height 7 miles.

For perimeter we need to apply the Pythagorean Theorem to the two right triangles and find their hypotenuses:

$$\text{Top hypotenuse } = \sqrt{7^2 + 1^2} = \sqrt{50} \approx 7.1 \text{ miles}$$
$$\text{Bottom hypotenuse } = \sqrt{7^2 + 3^2} = \sqrt{58} \approx 7.6 \text{ miles}$$
$$\text{Perimeter } \approx 6 + 7.1 + 10 + 7.6 = 30.8 \text{ miles.}$$

> ☞ A trapezoid is a shape with four sides, two of which are parallel (bases).
>
> **Area of a trapezoid**
> $\tfrac{1}{2}$(**base 1 + base 2**) × (**height**)

b) The shape can be decomposed into a 32×32 square and a 64×16 rectangle.

$$\text{Area } = (32 \text{ in})^2 + (64 \text{ in }) \times (16 \text{ in }) = (1024 \text{ in}^2) + (1024 \text{ in}^2) = 2048 \text{ in}^2.$$

If we add 3 sides of the square, 3 sides of the rectangle, and the *difference* between the length of the rectangle and the side of the square, we get:

$$\text{Perimeter } = 3 \times (32 \text{ in}) + (64 \text{ in}) + 2 \times (16 \text{ in}) + (64 \text{ in} - 32 \text{ in}) = 234 \text{ in.}$$

c) We take the *difference* between the rectangle and triangle areas:

$$\text{Area } = (6 \text{ cm}) \times (10 \text{ cm}) - \tfrac{1}{2}(4 \text{ cm }) \times (5 \text{ cm }) = 50 \text{ cm}^2.$$

The height of the triangle halves the base and thus, the triangle is *isosceles* with two sides of length

$$\text{Leg } = \sqrt{4^2 + (5/2)^2} = \sqrt{22.25} = 4.7 \text{ cm}$$
$$\text{Inner perimeter } = 2 \times \text{ Leg } + \text{ base } = 2 \times 4.7 + 5 = 14.4 \text{ cm}$$
$$\text{Outer perimeter } = 2 \times (6 + 10) = 32 \text{ cm.}$$
$$\text{Perimeter } = \text{ Inner } + \text{ Outer } = 14.4 + 32 = 46.2 \text{ cm.}$$

Perimeter and Area of a Circle

As another example, a *circle* is a shape formed by points that are equally apart from a fixed point called *center*. The distance between any of those points and the center is called the *radius* of the circle and twice the radius is the *diameter* of the circle. The perimeter of a *circle* is called *circumference* and it is a geometric fact that there is a constant $\pi \approx 3.14$ such that

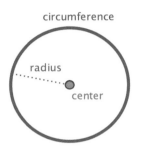

Figure 0.9 Circle

(0.11)   **Circumference** $= 2\pi \times$ **(radius)** $= \pi \times$ **(diameter)**

(0.12)   **Area of a circle** $= \pi \times$ **(radius)**^{2}

To check the 1st formula experimentally, we can use a piece of string wrapped around a circular shape and compare its length with the diameter of that shape. Notice that the diameter is a distance and the square radius is an area.

Example 0.3 Find the area and perimeter of the shape in **Fig 0.10** where the grid square is 1 sq. ft and the arc is a semicircle.

Solution: The center of the semicircle should be equally apart from all the points of the circle so that the radius is 3 feet. The shape can be decomposed into a 3 × 6 rectangle and a semicircle of radius 3 feet.

$$\text{Area} = (6\text{ ft})(3\text{ ft}) + \tfrac{1}{2}\pi(3\text{ ft})^2 \approx 32.14 \text{ sq. ft}$$
$$\text{Perimeter} = (3\text{ ft}) + (6\text{ ft}) + (3\text{ ft}) + \pi(3\text{ ft}) \approx 21.4 \text{ ft.}$$

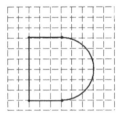

Figure 0.10 Where is the center?

Example 0.4 Find the area and perimeter of the shaded shape in **Fig 0.11** where the arc is a semicircle.

Solution: The diameter of the semicircle is twice its radius so that the length of the rectangle is 4 in. The area is the difference between the rectangle and the inner semicircle areas, but the perimeter is still additive.

$$\text{Area} = (4\text{ in})(3\text{ in}) - \tfrac{1}{2}\pi(2\text{ ft})^2 \approx 5.7 \text{ sq. in}$$
$$\text{Perimeter} = (3\text{ ft}) + (4\text{ ft}) + (3\text{ ft}) + \pi(2\text{ ft}) \approx 16.3 \text{ ft.}$$

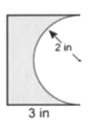

Figure 0.11 What is the diameter?

Volumes of Prisms and Cylinders

An *oblique prism* is a 3D shape (solid) generated by swiping a 2D *rectilinear* shape (polygon) called the *base* along a fixed direction such that each *vertex* of the base is tracing a fixed distance called the *lateral edge*. If the swipe direction is *perpendicular* to the base, then the prism is called a *right prism* and the lateral edge is called the *height* or *length* of the prism. For example, a unit square will swipe a *unit cube* if the swipe direction is perpendicular to the base such that the lateral edge is one unit long. These unit cubes are units of measure for *volume*: cubic inches, cubic feet, cubic miles, cubic centimeters, cubic meters, etc.

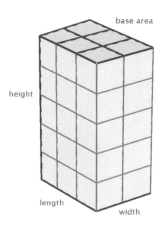

Figure 0.12 Volume counting unit cubes = **height × length × width**

> **Definition 0.5** *Volume*
>
> The **volume** of a three dimensional shape is the *exact* number of unit cubes that fit inside that shape. A volume could be any positive real number.

In particular, each unit square in the base of a right prism will generate one unit cube for each unit of its height. Since the number of square units in the base is the base area, this leads to

(0.13)   **Volume of a (right) prism = (height) × (area of the base).**

For example, if the box in **Fig 0.12** has the height 5 ft, the length 3 ft, and the width 2 ft, then the base area has 3 ft × 2 ft or 6 sq. ft, and each sq. ft in the base will generate 5 cubic ft along the height. The volume is

Volume of the box = (5 ft) × (3 ft) × (2 ft) = (5 ft) × (6 sq. ft) = 30 cubic ft.

If the base is a circle, then the same procedure will produce an *oblique cylinder*. The center of the circle will trace the *axis* of the cylinder with the same length as the *lateral edge*. A cylinder is a *right cylinder* if its axis is perpendicular to the base. In this case, the *height* or *length* of the cylinder is the length of its axis.

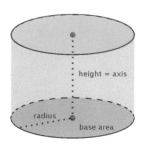

Figure 0.13 Right cylinder

(0.14)   **Volume of a (right) cylinder = (height) × π (radius)^2.**

This formula is a special case of the volume of the prism formula as the area of the base for a cylinder is $\pi(\text{radius})^2$.

Example 0.5 Find the right cylinder volume **Fig 0.14** with diameter 18 in.

Solution: The radius is half of the diameter, radius = 18/2 = 9 in, and the length of the cylinder is 11 in.

$$\text{Volume} = (11 \text{ in }) \times \pi(9 \text{ in })^2 \approx 2,799 \text{ in}^3.$$

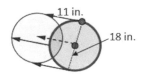

Figure 0.14 What is the radius?

Example 0.6 Find the right prism volume **Fig 0.15** with right triangle base.

Solution: To find the area of the base we need to find its height. Using Pythagoren Theorem we have height $= \sqrt{109^2 - 91^2} = \sqrt{3600} = 60$ ft and

$$\text{Volume} = (74 \text{ ft}) \times \left(\tfrac{1}{2}(60 \text{ ft}) \times (91 \text{ ft})\right) = 202,020 \text{ ft}^3.$$

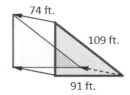

Figure 0.15 What is the height?

Wrapping Up

We have seen perimeter and area of polygons and circles as well as volumes of prisms and cylinders. These will be used in problems throughout algebra. The following formulas assume *common units* of length.

Perimeter: the length (exact number of units) of the boundary of the shape.

(rectangle)   Perimeter of a rectangle = $2 \times$ (length + width)

(circle)   Perimeter of a circle = $2 \times \pi$ (radius)

(semicircle)   Perimeter of a semicircle = $\pi \times$ (radius)

Pythagorean Theorem

(right triangle)   (hypotenuse)^2 = (leg one)^2 + (leg two)^2

Area: the exact number of unit squares that fit inside the shape.

(rectangle)   Area of a rectangle = (length) × (width)

(square)   Area of a square = (side length)^2

(parallelogram)   Area of a parallelogram = (height) × (base)

(trapezoid)   Area of a trapezoid = (height) $\times \frac{1}{2}$ (base one + base two)

(triangle)   Area of a triangle = $\frac{1}{2}$ (height) × (base)

(circle)   Area of a circle = π (radius)^2

Volume: the exact number of unit cubes that fit inside the shape.

(prism)   Volume of a prism = (height) × (area of the base)

(box)   Volume of a rectangular prism (box) = (height) × (length) × (width)

(cube)   Volume of a cube = (side length)^3

(cylinder)   Volume of a cylinder = (height) $\times \pi$ (radius)^2

Exercises

Exercises for 0.2 Perimeter, Area, and Volume

P0.6 Calculate the perimeter & area of the figure on the grid if each square represents 1 square mile.

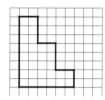

P0.7 Calculate the perimeter & area of the figure on the grid if each square represents 1 square foot.

P0.8 Calculate the perimeter & area of each figure on the grid if each square represents 1 square cm.

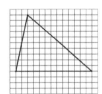

a)

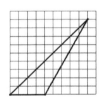

b)

P0.9 Calculate the perimeter & area of the figure on the grid if each square represents 1 sq. meter.

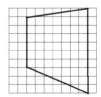

P0.10 Calculate the area of the figure.

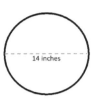

14 inches

P0.11 Calculate the perimeter of the figure.

side length = 23 cm

equilateral triangle

P0.12 Calculate the perimeter & area of each figure.

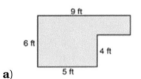

a)

9 ft
6 ft
4 ft
5 ft

b)

32 in
32 in
16 in
64 in

c) The right end is an isosceles triangle.

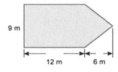

9 m
12 m   6 m

P0.13 Calculate the perimeter & area of each figure where the arcs are semicircles.

2 in
3 in

a)

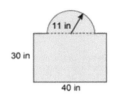

b)

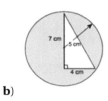

b)

P0.14   Calculate the perimeter and area of each figure on the grid where all arcs are semicircles.

    a) Each square is 1 square meter.

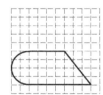

    b) Each square is 1 square mile.

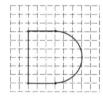

    c) Each square is 1 square inch.

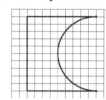

P0.15   Calculate the inner perimeter, the outer perimeter, and the area of each shaded region.

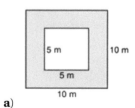

a)

P0.16   Calculate the volume of each right prism or cylinder shown below (not drawn to scale).

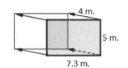

a)

b) This is a cube.

c) The diameter is 18 in.

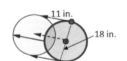

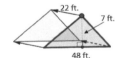

d)

e)

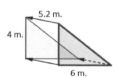

f)

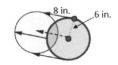

g)

Unit 1

Variables and Functions

1.1   Variable versus Constant Quantities

Launch Exploration

Suppose that you have the following two situations:

(**1.1**) Last week you went shopping and spent exactly $180 on shirts and jeans. The shirts cost $15 each and the jeans cost $30 each.

(**1.2**) Simone is going to fence in a rectangular pen in the yard for her dog to play in. The area of the pen will be 360 square feet. The fencing costs $12 per foot to install.

In order to understand each situation you need to ask questions about the quantities involved. For each situation, get together with a group of classmates and try to answer the questions in **Table 1.1** on the margin .

Analyzing a Situation: Quantities and Units

A quantity is an **amount** or **a type of amount** that can be counted or measured. In a real-world situation, the quantities involved are a natural part of the situation; they are *not a choice*. A unit or **unit of measure** is the **standard of measure** in which we *choose* to express amounts of a given quantity.

> **Fact 1.1   *Quantities & Units***
>
> A **quantity** (type of amount) and your choice of **unit** (standard of measure for the amount) are answers to: *"How much what?"* or *"How many what?"*

Before you move on, make a chart listing as many types of quantities as you can in one column, and possible choices of units for each quantity in the other.

Key Questions

Identify as many quantities related to this situation as you can.

Which quantities are known?

Which quantities are unknown?

What units could you use to express them?

Table 1.1

Examples of Quantities & Units

Time	seconds [sec], hours [hrs], days, weeks [wks], months [mos], years [yrs], milliseconds [msec], nanoseconds [nsec], centuries [cen]
Distance	inches [in], feet [ft], yards [yrds], miles [mi], millimeters [mm], centimeters [cm], decimeters [dm], meters [m], kilometers [km], light-years [ly]
Area	square inches [in^2], square feet [ft^2], square yards [yd^2], square miles [mi^2], square centimeters [cm^2], square meters [m^2], square kilometers [km^2], acres, hectares [ha]
Volume (solid)	cubic inches [in^3], cubic feet [ft^3], cubic yards [yd^3], cubic miles [mi^3], cubic centimeters [cm^3], cubic meters [m^3]
Volume (liquid)	pints [pt], gallons [gal], quarts [qt], barrels [bbl], fluid ounces [fl oz], liters [L]
Money	American dollars [$] or cents [¢], British pounds [£], Euros [€], Mexican pesos, Cuban pesos, Norwegian krones, Kenyan shillings, Korean wons
Power	watts [W], joules per second [J/s], horsepower [hp]
Work	words typed, rooms painted, items scanned, boxes assembled, cars detailed, lawns mowed, square feet mowed, pints of berries picked
Speed	feet per second [ft/s], feet per minute [ft/min], inches per second [in/s], miles per hour [mph], kilometers per hour [km/h], miles per day [mpd]
Rate of Work	words typed per minute [word/min], square feet painted per hour [ft^2/h], gallons filled per minute [gal/min], baskets weaved per week [bsk/wk], boxes assembled per hour
Other Rates	dollars earned per year, $ paid in tax per $ in price, $ in total cost per number of items

Analyzing a Situation: Variables and Constants

When we analyze a situation quantitatively, some quantities are unknown and we want to assign them variables that we can manipulate using algebra. Hence, it is helpful and important to first identify the relevant quantities as variables and constants and to choose their units of measure.

☞ Whether we represent a quantity by a variable or not depends on the context.

Definition 1.2 *Variables & Constants*

A **variable** is (a placeholder representing) a quantity that is unknown or that can vary throughout a situation.

A **constant** is (a number representing) a quantity that is known or that does not vary throughout a situation.

Choosing smart labels

As placeholders, variables are usually *labeled* by letters. Sometimes we can also use abbreviations or symbols. It is recommended to choose smart labels that help us remember what quantity the variable represents. For example, the total pay, the hourly rate of pay, and the hours worked can be labeled by the letters p, r, h or by the abbreviations pay, $rate$, hrs.

Numbers can be used as **values** of a variable. For example, if t is a variable representing time in hours and 2 is a number of hours, we say that *the variable t takes on the value* 2 and write $t = 2$ to show that we will substitute t with 2. However, do not be tempted to think that the variable t in this case becomes a constant! The correct interpretation is that the variable t takes on the value 2 and may very well take on other values, too.

Fact 1.3 *Values of a Variable*

As a placeholder, a (numerical) variable can **take on values** that are (real) numbers. These numbers do not change the nature of the quantity as a variable.

Continuous variables can take on *any* value between two specified values. All 'in-between' values make sense. Continuous variables represent quantities that are *measured*.

Example 1.1 The amount of water in a 6-ounce cup is a continuous variable. We *measure* the amount of water and *any* amount between 0 and 6 ounces does make sense (e.g. 0.125845 oz.) See **Fig 1.1**.

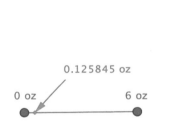

Figure 1.1 Continuous Variable

Discrete variables are not continuous - they only take on distinct, separate values. Not all 'in-between' values make sense. Discrete variables represent quantities that are *counted*.

Example 1.2 The cost of your grocery bill is a discrete variable. We *count* the amount of money, and *not all* values in-between $24.50 and $24.51 make sense (e.g. $24.502.) See **Fig 1.2**.

Figure 1.2 Discrete Variable

Definition 1.4 *Continuous vs. Discrete*

A variable is **continuous** if it can take on any value between two specified values and **discrete** if it takes on distinct, separate values.

☞ Constants are neither continuous nor discrete since they "don't take on values."

For each situation in **Example 1.3** and **Example 1.4** make two lists: a) a list of variables; b) a list of constants. Identify units for all of the quantities in each list. Compare your lists and your initial answers in the exploration with those of other classmates. Determine whether each variable is continuous or discrete.

Example 1.3

Last week you went shopping and spent exactly $180 on shirts and jeans. The shirts cost $15 each and the jeans cost $30 each.

Solution: If your constants were $180, $15, and $30, you are on the right track, but you should pay attention to the phrase "cost ... each". So while 180 is the total cost of shirts and jeans in $, 15 and 30 are actually *rates* measured respectively in $ *per* shirt and $ *per* jean. The unknown quantities are the number of shirts and the number of jeans counted as separate items. That means that we have two discrete variables that we can conveniently label by the first letters: s shirts and j jeans. We organize this information in the margin.

The next example involves some geometry.

Example 1.4

Simone is going to fence in a rectangular pen in the yard for her dog. The area of the pen will be 360 square feet. The fencing costs $12 per foot to install.

Solution: The constants are 360 and 12 where 360 is the pen area measured in ft^2 and 12 is the cost of fencing per foot and thus, it is a rate measured in $/ft. The unknown quantities (variables) are the width and the length of the rectangle labeled respectively by the letters w and l, the perimeter labeled by p, and the total cost of fencing labeled by C. Since *any* value between two given values of length, say between 4 ft and 5 ft makes sense, we conclude that the variables w and l are continuous. Since we *count* the money for the total cost, C is a discrete variable. This information is organized in the margin.

To extend your creativity, do **Exercise P1.2**.

Values of a Variable - Discussion

Variables are placeholders for quantities with unknown values, and the number of values that a variable can take on depends on the situation. When we assign variables we do not *yet* know their values - finding their values requires additional work. Let's look at two different cases.

A variable may take on infinitely many values

Suppose a rectangle has an area 36 ft^2 (square feet). In this situation, the width and the length of the rectangle can change throughout the situation. For example,

---

Answer Example 1.3

a) Variables
  s (shirts) items, discrete
  j (jeans) items, discrete

b) Constants
  total cost 180 $
  cost per shirt 15 $/shirt
  cost per jean 30 $/jean

Answer Example 1.4

a) Variables
  w (width) ft, continuous
  l (length) ft, continuous
  p (perimeter) ft, continuous
  C (total cost) $, discrete

b) Constants
  pen area 360 ft^2
  cost of fencing per foot 12 $/ft

if we label the width of the rectangle by w and the length by l, we can have $w = 12$ ft (feet) and $l = 3$ ft (feet), or we can have $w = 8$ ft and $l = 4.5$ ft. Indeed, in both cases the area is $12 \times 3 = 8 \times 4.5 = 36$ ft^2. In fact, w and l are continuous variables that can take on *infinitely many values*.

☞ Area of a rectangle equals the length times the width.

A variable might take on a single value

Suppose now that a *square* has an area of 36 ft^2. In this situation, the side length of the square is *unknown* and thus, it can be represented by a *variable*, say s. However, the only possible value for this variable is $s = 6$ ft since this is the only side length for which the area of the square is $6^2 = 36$ ft^2. Hence, the variable s has only *a single value in the situation*. Notice that the variable s is still *continuous* since as a length it can take on *any* value between two specified (hypothetical) values.

☞ Since a square has four sides of equal length, only one variable for side-length is needed.

So depending on the situation, a variable might take on two or more different values (the values can *vary*), it might have only one possible value, or there might not be any valid value for the variable. Can you find such a situation? We will see more examples of each of these types of situations throughout this course. [1]

Wrapping Up

Algebra is a language that helps us to answer questions and to solve problems in the world around us. The key to understanding and using this language is to be able to identify and classify the quantities in a situation. Hence, the first steps in solving any problem with algebra should always be to identify the relevant quantities, determine whether each is a variable or a constant, choose and iden-tify letters (or abbreviations) to represent the variables, determine whether the variables are continuous or discrete quantities, and to choose units to measure each quantity. Keep these first key steps in mind and you will be on your way to a better understanding of how to use Algebra!

You can use the flowchart in **Fig 1.3** as a logical scheme to help you. To check your understanding do **Exercise P1.1**.

First Keys to Algebra

Identify the relevant quantities in a situation.

Determine whether each is a variable or a constant.

Label the variables by letters or abbreviations.

Determine whether the variables are continuous or discrete.

Choose units to express each quantity.

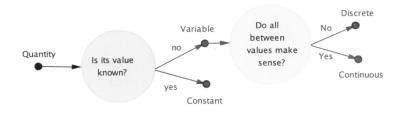

Figure 1.3 Classifying the Quantities.

Exercises

Exercises for 1.1 Variable versus Constant Quantities

P1.1   Explain to a classmate the meaning of each word in the following vocabulary list: *quantity, unit of measure, variable, constant, continuous variable, discrete variable.* Write a definition or explanation for each in your own words and create your own examples.

P1.2   For each situation, make two lists - one list of variables in the situation, and one list of constants in the situation. Determine whether each variable is continuous or discrete. Identify units for all of the quantities in each list.

    a) You are filling your car's gas tank.

    b) You travel by car from Hartford to Atlanta.

    c) You make several credit card purchases.

    d) You are scuba diving.

    e) You are draining the water in your bathtub.

Get creative! We are not trying to solve a problem or to answer a specific question. Think outside the box about quantities that could be related in *any way* to these situations.

P1.3   The population of a town was 12,300 at one point and it has been decreasing by 2.3% each year. Which of the following are constants in this situation?

    a) The total population

    b) The starting population

    c) The time that has passed since the population was 12,300

    d) The population's rate of decrease

P1.4   A handywoman charges $20 to come to your house to give an estimate for a job and then another $35 for every hour she works on the job. Which of the following are variables in this situation?

    a) The total cost of the job

    b) The cost of the estimate

    c) The amount of time the handywoman works on the job

    d) The hourly rate the handywoman charges

P1.5   Marni bought two kinds of nuts - cashews and almonds. She spent $42, and for that money she bought 2 pounds of cashews and 3 pounds of almonds. Which of the following are constants in this situation?

    a) The cost of cashews ($ per pound)

    b) The cost of almonds ($ per pound)

    c) The total cost ($)

    d) The total amount of nuts purchased (pounds)

P1.6   The width of a rectangular mirror is 12 cm fewer than its height. The mirror weighs 6.2 pounds and measures 168 centimeters around the outside. Which of the following are constants in this situation?

    a) The perimeter of the mirror

    b) The width of the mirror

    c) The weight of the mirror

    d) The height of the mirror

P1.7   Zoe is going to build a rectangular fenced-in pen for her dog along one side of her house, so she only needs to fence three sides. She has 160 feet of fencing and wants to maximize the area inside the pen. Choose the answer that BEST describes: the area inside the pen in this situation.

    a) a constant - units: feet

    b) a variable - units: square feet

    c) a variable - units: feet

    d) a constant - units: square feet

P1.8   Daria bought burgers and chicken sandwiches for the team. She bought 30 burgers and 20 chicken sandwiches and she spent $120 all together. Choose the answer that BEST describes: the value of 120 in this situation

    a) a variable - total amount of food (burgers and sandwiches)

    b) a constant - cost (dollars per burger and per sandwich - a rate)

    c) a variable - total cost (dollars)

d) a constant - total amount of food (burgers and sandwiches)

e) a variable - cost (dollars per burger and per sandwich - a rate)

f) a constant - total cost (dollars)

P1.9 Last week you went shopping and spent exactly $180 on shirts and jeans. The shirts cost $15 each and the jeans cost $30 each. Choose the answer that BEST describes: the value of 30 in this situation.

a) a variable - cost (dollars)

b) a variable - cost (dollars per pair of jeans - a rate)

c) a constant - amount of jeans (jeans)

d) a variable - amount of jeans (jeans)

e) a constant - cost (dollars)

f) a constant - cost (dollars per pair of jeans - a rate)

P1.10 Daria bought burgers and chicken sandwiches for the team. She bought 30 burgers and 20 chicken sandwiches and she spent $120 all together. Choose the answer that BEST describes: the cost of a burger in this situation

a) a constant - units: dollars

b) a variable - units: dollars per burger - a rate

c) a constant - units: dollars per burger - a rate

d) a variable - units: dollars

P1.11 Determine whether the variable is continuous or discrete.

a) A dog's weight over its lifetime.

b) The area of a back yard.

c) The number of classes you take in a semester.

d) A person's height over their lifetime.

e) The number of socks in a drawer.

f) The amount of time a person is driving on a trip.

g) The speed of a plane.

h) The amount of water in a pitcher.

P1.12 A scientist starts with 500 bacteria in a petri dish. The bacteria population increases by 1/5 every week. Which of the following are variables in this situation?

a) The bacteria population's rate of growth

b) The starting amount of bacteria

c) The amount of time since the bacteria population has been growing in the dish

d) The total number of bacteria in the petri dish at any time

P1.13 Paola invested $2,000 in an account that pays 1.1% interest compounded weekly. Which of the following are variables in this situation?

a) The annual interest rate that the bank pays

b) The amount of time since Paola opened the account

c) The starting amount of money in the account ($)

d) The amount of money in the account ($) at any time

P1.14 Sam started saving money for a down payment on a car. He first put $460 in his savings envelope and then added $60 every week. Choose the answer that BEST describes: the amount of time Sam has been saving money.

a) a constant - units: weeks

b) a variable - units: weeks

c) a variable - units: dollars per week (a rate)

d) a constant - units: dollars per week (a rate)

1.2   Translate English to Algebra and Back

Launch Exploration

Suppose we are building a rectangular deck. We want the area of the deck to be at least 200 square feet but we use exactly 960 inches of trim for the edges. We also want the length of the deck to be five feet less than twice its width.
Read this information carefully and do the tasks listed in **Table 1.2** in the order that makes sense to you.

Operations and Relations - Key Phrases

In a verbal description of a situation, **operations** and **relations** between variables and constants are expressed via **key phrases** and the way these are put together. The result of a basic translation could be a symbolic expression built up from variables and constants by operations (*algebraic expression*) or a relation between two quantities which are either equal (*equation*) or one of which is greater than the other (*inequality*). Below are a few generic examples of basic translations from English into symbolic Algebra involving a single variable.

A Few Examples of Basic Translations for Reference

Expression in English	Algebraic Expression
A number *plus* ten	$a + 10$
Six *more than* a number	$b + 6$
The sum of a number and two	$c + 2$
The total of eight and some number	$8 + d$
A number *increased by* eleven	$e + 11$
Fifteen *added to* a number	$f + 15$
A number *minus* fifteen	$g - 15$
Eight *less than* a number	$h - 8$
The difference of a number and six	$i - 6$
A number *decreased by* ten	$k - 10$
A number *subtracted from* three	$3 - l$
Four *times* a number	$4m$
The product of eighteen and a number	$18n$
Twice a number; *double* a number	$2p$
A number *multiplied by* negative fifty	$-50q$
Seven tenths *of* a number	$\frac{7}{10}r$
The quotient of a number and seven	$\frac{s}{7}$ or $s \div 7$
Negative forty *divided by* a number	$\frac{-40}{t}$ or $-40 \div t$
The ratio of a number to thirteen	$\frac{v}{13}$

François Viète (1540-1603, French) was born in Fontenay-le-Comte (now Vendée), France. He was the first mathematician who introduced notations for knowns and unknowns in solving a problem and replaced procedures by symbolic algebra. His approach to problem solving was to form equations, identify their types, and solve them in the context of the situation. (Wikipedia)

The square of a number; a number *squared*	u^2
The cube of a number; a number *cubed*	w^3
A number *to* the second *power*	x^2
A number *to* the third *power*	y^3

Requirement/ Constraint in English	Equation
When three is added to a number the result is ten.	$x + 3 = 10$
Five less than a number *equals* fifteen.	$z - 5 = 15$
Negative six more than a number *is* eighteen.	$-6 + A = 18$
Negative forty-four *is the same as* three times a number.	$-44 = 3B$
The difference of a number and nine *yields* twenty.	$C - 9 = 20$
The quotient of a number and ten *amounts to* sixty.	$\frac{D}{10} = 60$

Requirement/ Constraint in English	Inequality
A number *is at least* five.	$E \geq 5$
A number *is at most* three quarters.	$F \leq \frac{3}{4}$
Eight *is less than* a number; *smaller than.*	$8 < G$
A number *is more than* ten; *greater than.*	$H > 10$
Any number *is* either *positive* or *nonpositive.*	$I > 0$ or $I \leq 0$
Twice a number *is no more than* nine.	$2J \leq 9$
Any number *is* either *nonnegative* or *negative.*	$K \geq 0$ or $K < 0$
The sum of five and triple a number *is up to* thirty.	$5 + 3L \leq 30$

☞ **Equivalent Translations**

Notice that $b + 6 = 6 + b$ and $E \geq 5$ means $5 \leq E$. Can you find other expressions or relations equivalent to the ones in the list?

☞ **Strict vs. Non-strict Ineq.**

The inequality $a < 0$ (negative) is *strict* since it excludes $a = 0$. The inequality $a \leq 0$ (nonpositive) is *non-strict* since it includes $a = 0$.

Algebraic operations include the four arithmetic operations, *addition, subtraction, multiplication,* and *division,* as well as *taking powers a^n, roots $\sqrt[n]{a} = a^{1/n}$,* and *reciprocals* of non-zero numbers $a^{-n} = \left(\frac{1}{a}\right)^n$ with n a positive integer. (Defining a^n for n irrational is beyond the scope of Algebra.)

Definition 1.5 *Algebraic Expressions vs. Equations & Inequalities*

An **algebraic expression** is a combination of variables and constants connected by *algebraic operations*. An **equation** is an equality (relation) between two expressions involving one or more variables. An **inequality** is a relation between two quantities of which one is greater than the other.

An algebraic expression can be *simplified* by algebraic manipulations or *evaluated* at admissible values of the variables, but *cannot be solved* since it is not a relation. An equation can be *solved* by finding all values for the variables that make it a true statement, meaning that the evaluated expressions on both sides of the equal sign agree. In this case, the values for the variables are called *solutions*

☞ Equations and inequalities are read from LEFT to RIGHT:

$a = b$ reads a equals b

$a < b$, a is less than b

$b > a$, b is greater than a

$a \leq b$ reads $a = b$ **or** $a < b$

$b \geq a$ reads $b = a$ **or** $b > a$

$a < x < b$, $a < x$ **and** $x < b$

to the equation. These could be single values, pairs, triples, etc. depending on how many variables are involved. This terminology applies to an inequality by replacing the equal sign with less than (<) or greater than (>).

☞ Expressions encode operations; equations and inequalities are relations between expressions.

Definition 1.6 *Solutions to Equations & Inequalities*

An equation or inequality can be a **true** or a **false** statement depending on the values of the variables. If it is a true statement, the values of the variables are called **solutions** to the equation or inequality. **Solving** an equation or inequality means finding all solutions.

Algebraic Expressions vs. Equations & Inequalities

For each example we 1) identify it as an algebraic expression, equation, or inequality; 2) give an appropriate instruction to execute; 3) determine the outcome.

Symbolic	Type	Instruction	Outcome
a) $x^2 + 1/x$	Expression	Evaluate at $x = 1/3$	$28/9$
b) $-7y + 4 = 0$	Equation	Solve for y	$y = 4/7$
c) $-7z + 4 < 0$	Inequality	Is $z = -4$ a solution?	No
d) $s^2 + t^2$	Expression	Evaluate at $s = -1$ and $t = 2$	5
e) $-7x + 4y \geq 0$	Inequality	Is $(x, y) = (4, 7)$ a solution?	Yes
f) $z^2 = 25 - p^2$	Equation	Is $(p, z) = (-4, 3)$ a solution?	Yes
g) $2(u - t^2) + 3t^2$	Expression	Simplify completely	$2u + t^2$

a) No statement, so it is an expression. We can evaluate it by replacing x with 1/3 in $x^2 + 1/x$ and follow the order of operations. The outcome is a numerical value: $(1/3)^2 + 3 = 1/9 + 27/9 = 28/9$.

b) This is a linear equation which we can solve by isolating the variable y. Add $7y$ to both sides, $4 = 7y$; divide both sides by 7; get the only solution $y = 4/7$. To verify it, replace y by 4/7 in $-7y + 4$ and check the statement $-7(4/7) + 4 = 0$ is true: $0 = 0$.

c) This is an inequality and to check whether $z = -4$ is a solution, we replace z by -4 in $-7z + 4$ and ask whether the statement $-7(-4) + 4 < 0$ is true. The answer is no since by the order of operations $-7(-4) + 4 = 28 + 4$ is positive, not negative (< 0.)

d) This is an expression which can be evaluated at $(s, t) = (-1, 2)$ by replacing s by -1 and t by 2 in $s^2 + t^2$ and watch out for the order of operations. The numerical value of the expression is $(-1)^2 + (2)^2 = 1 + 4 = 5$.

e) This is an inequality. The pair $(4, 7)$ is a solution since by replacing x with 4 and y with 7 in $-7x + 4y$ we get a true statement: $-7(4) + 4(7) \leq 0$ ($0 = 0$ is included.)

f) This is an equation. The pair $(-4, 3)$ is a solution since it makes the equality a true statement: $z^2 = (3)^2 = 9$ agreeswith $25 - p^2 = 25 - (-4)^2 = 25 - 16 = 9$.

☞ The reciprocal of $\frac{1}{3}$ is 3; the square of $\frac{1}{3}$ is $\left(\frac{1}{3}\right)\left(\frac{1}{3}\right) = \frac{1^2}{3^2} = \frac{1}{9}$.

g) This is an expression we can simplify; distribute the factor 2 and group like terms:

$$2(u - t^2) + 3t^2 = 2u - 2t^2 + 3t^2$$
$$= 2u + (3t^2 - 2t^2)$$
$$= 2u + t^2.$$

Relations in Situations

Going back to the **Launch Exploration**, the word "rectangular" tells us that the deck's outline is a *rectangle*. So, there are two pairs of opposite sides having the same distance; the longer one is called the *length* while the shorter one is called the *width* of the rectangle. "The edges" of the deck are the sides of the rectangle and in this case, "the trim" is the *perimeter* of the rectangle. We identify four variables: the length ℓ, the width w, the perimeter p of the rectangle all in feet, and the area A of the rectangle in square feet. See **Fig 1.4**. The *main constants* are 200 square feet of area, and 960 inches converted to 80 feet of trim. To find the relations among the variables and constants, we need to answer two questions:

1) *What are the given requirements/constraints? Translate them.*

The area is *at least* 200 square feet:	$A \geq 200$
We *have only* 80 feet of trim:	$p = 80$
The length *is* five feet *less than twice* the width:	$\ell = 2w - 5$

Notice that the constants 5 and 2 are embedded in the description as "five feet less than" and "twice".

2) *What other information is related to the situation? Write related formulas.*

The area of a rectangle *is* its length *times* its width:	$A = \ell \times w$
The perimeter of a rectangle *is* the *sum* of its side lengths:	$p = 2\ell + 2w$

The added relations are not explicitly given in the verbal description, but are formulas from geometry related to rectangles. However, not every formula about a rectangle is *related* to the situation. For example, the sum of the squares of a rectangle's length and width is always equal to the square of the length of its diagonal (Pythagoras), but the diagonal is not mentioned in our situation.

Definition 1.7 *Formulas*

A **formula** is like a recipe, it shows a sequence of operations to be performed on certain quantities in order to obtain another quantity.

A formula can be expressed verbally as a step by step procedure (algorithm) or symbolically by an (algebraic) expression or by an equation of the form

quantity = expression in terms of other quantities.

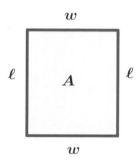

Figure 1.4 $p = \ell + w + \ell + w$.

☞ Whether a formula is related or not depends also on the question.

☞ Express formulas both verbally and symbolically!

It is important to use the labels we chose for our variables *consistently*. For example, the formula for the area of a rectangle could be given as $S = x \times y$ where x is the length and y is the width. Translating this formula into English shows how to match its labels with our labels: $S = A$, $x = \ell$, and $y = w$. Expressing formulas both verbally and symbolically helps us better understand how to use them.

Often the best way to communicate math is to combine English words with symbols. A good way to check your mathematical writing is to read it out loud including the symbols. It should make sense to you and to your audience.

☞ Don't forget to pay attention to English grammar and Order of Operations! The word *quantity* means a grouping of operations.

Reading Symbolic Algebra in English

Symbolic Algebra	English
-2^3	the opposite of the cube power of 2; NOT the cube power of negative 2 (expression)
$5x - 6y = 7$	Six times a number subtracted from five times another number equals seven. (not a formula)
$\frac{-b + \sqrt{b^2 - 4ac}}{2a}$	the opposite of b plus the positive square root of *quantity* b squared minus $4ac$, all divided by $2a$ (expression)
$V = \ell \cdot w \cdot h$	The volume of a box is the product of its length, width and height. (volume formula)
$u(u - 3) \le 200$	A number times *quantity* 3 less than that number is at most 200. (not a formula)
$A = \frac{h}{2}(b_1 + b_2)$	The area of a trapezoid is the product of half the altitude and the sum of the bases. (area formula)

Analyzing Situations - More Examples

In each situation below do the the following tasks:

- Identify and label the variables and constants including their units.
- Translate the requirements and constraints into equations and inequalities.
- Write any additional formulas that are related to the situation.

Whenever appropriate, draw a diagram and label it with variables and constants.

a) *Sergio believes he is five years younger than double the age of Chloe, and Chloe believes she is five years older than half the age of Sergio.*

Solution: *Variables:* Let S be Sergio's age in years. Let C be Chloe's age in years.

Requirements and constraints:

Sergio *is* five years *younger than double* the age of Chloe: $S = 2C - 5$

Chloe *is* five years *older than half* the age of Sergio: $C = \frac{1}{2}S + 5$

The constants 5, 2 and 1/2 are embedded in the description as "five", "double", and "half".

Related formulas: None so far.

b) *Abe, Bella, and Carlos have all earned some college credits. Bella has earned 17 credits more than Abe, and Carlos has earned 9 less than triple the number of Abe's credits. Altogether, they have earned a total of 183 credits.*

Solution: *Variables:* Let A be the number of credits that Abe has earned. Let B be the number of credits that Bella has earned. Let C be the number of credits that Carlos has earned.

Main Constant: 183 credits total.

Requirements and constraints:

Bella *has* 17 credits *more than* Abe: $B = A + 17$

Carlos *has* 9 *less than triple* the number of Abe's: $C = 3A - 9$

Altogether, they have 183 credits: $A + B + C = 183$

The constants 17, 9, and 3 are embedded in the description as "17 credits", "9 credits", and "triple."

Related formulas: None so far.

c) *Mateo has 14 meters of fence for a rectangular pen alongside his house. He only needs to fence three sides such that the length is 1 meter less than triple the width. The pen is for his dog who needs more than 20 square meters to be happy.*

Solution: See **Fig 1.5** and **Fig 1.6**.

Variables: ℓ the length of the pen in meters; w the width of the pen in meters; f the length of the fence in meters; A the area of the pen in square meters.

Main constants: 14 meters of fence; 20 square meters of area.

Requirements and constraints:

Mateo has 14 meters of fence: $f = 14$

The length *is* 1 meter *less than triple* the width: $\ell = 3w - 1$

His dog needs *more than* 20 *square meters*: $A > 20$

Notice that the constants 3 and 1 are embedded in the description as "1 meter" and "triple." Also we use the unit *square meter* as a key phrase for the area.

Related formulas:

The area of a rectangle *is* its length *times* its width: $A = \ell \times w$

He needs to fence only *three* sides: $f = \ell + 2w$ or $f = 2\ell + w$

The pen is *alongside* the house and so there is no fence on that side. However we don't know if that side length is ℓ as in **Fig 1.5** or w as in **Fig 1.6** and we should exercise our critical thinking and consider *both* cases.

house

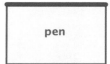

pen

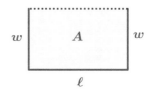

w A w

ℓ

Figure 1.5 $f = w + \ell + w$

house

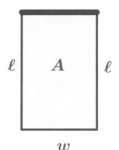

ℓ A ℓ

w

Figure 1.6 $f = \ell + w + \ell$

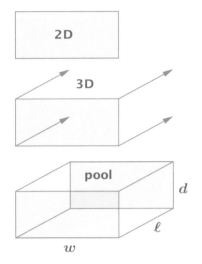

Figure 1.7 Building a Pool

d) *Pam has a box-shaped lap pool that is two times wider than it is deep and six times longer than it is wide. The water charge to fill the pool is $6 per 100 cubic feet and she wants to pay no more than $100.*

Solution: See **Fig 1.7**.

Variables: the width w, the length ℓ, and the depth d of the pool in feet; V the volume of the pool in cubic feet; C the total cost to fill in the pool in dollars.

Main Constants: $6 per 100 cubic feet (rate); $100 the maximum total cost.

Requirements and constraints:

The pool is *two times wider than* it is deep : $w = 2d$

The pool is *six times longer than* it is wide: $\ell = 6w$

She wants to pay *no more than* $100: $C \le 100$

The constants 2 and 6 are embedded as "two times" and "six times".

Related formulas:

The total cost *is* the rate *times* the volume: $C = \frac{6}{100}V$

The volume of a box *is* the *product* of its width, length, and depth: $V = w\ell d$

Units Agreement

When writing formulas or equations, units must agree. For example, given a rectangle of width 1 meter and length 2 feet, we must choose a common unit to find its area. We either convert the width of 1 meter to 3.28 feet or the length of 2 feet to 0.61 meters. So the area is either about 6.56 square feet or the area is about 0.61 square meters. Here is an example of disaster that resulted from a failure to make a unit conversion: "NASA lost $125 million Mars orbiter because a Lockheed Martin engineering team used English units of measurement while the agency's team used the [...] metric system." (CNN, Sep. 30, 1999)

Wrapping Up

Translations play an important part in understanding a problem. Given a situation verbally we translate English to math by using key phrases and the way they are arranged in the description. The outcomes of basic translations are algebraic expressions built up from variables and constants by operations and equations or inequalities which are relations between expressions. Expressions can be simplified by algebraic manipulations or can be evaluated to get a numerical value. Equations can be solved to find numerical solutions that make the equations true statements. In addition to the direct translations, certain formulas from geometry, economics, science etc. provide more relations among the variables and the constants in a situation. A formula is a 'recipe' used to get one quantity in terms of others. The tasks listed in **Table 1.7** will help us understand a problem and devise a plan to solve it in the next section (Pòlya.)

Keys to Analyzing a Situation

Identify and label the variables & constants including their units.

Whenever appropriate, draw a diagram and label it with variables and constants.

Translate the requirements and constraints into equations and inequalities.

Write any additional formulas that are related to the situation.

Table 1.7

Exercises

Exercises for 1.2 Translate English to Algebra and Back

P1.15 Translate from English into algebra. Identify and label your variables before you write your translation.

a) The sum of the length and 5 meters

b) A number subtracted from 20

c) The product of 7 and the sum of the radius and 2.

d) 15 square inches less than triple the area

e) $ 10 decreased by quantity $1 more than 5 times the price

f) The cost increased by 40 cents is equal to $4.50.

g) 8 less than triple the volume is at least 110.

h) The square of the opposite of the distance

i) 30 feet subtracted from 9 times the width

j) Eight cm less than six times the length is 172 cm.

k) The ratio of height to weight

l) The average of her 3 quizzes, is five more than the average of her two tests.

m) The quotient of a number and 8 is at most 10.

n) 52 increased by quantity 4 times the perimeter and 2 is 91 meters.

o) The sum of the time and 29 min. is more than 73.

p) The opposite of the square of the distance

q) Twice the sum of a circle's radius and a square's side length is no more than 100 feet.

r) The sum of twice a circle's radius and a square's side length is 100 feet.

s) The principal square root of the sum of two numbers.

t) The negative square root of the difference between the squares of two numbers

u) The product of the difference between in the two ages and four is twice the sum of the two ages.

v) The square of the difference between x1 and x2

w) The sum of four and twice the width is 36.

x) The square of the quotient of a number and 3 is 64.

y) Four times the average of 3 students' ages

z) Twice the sum of height and weight

P1.16 Translate into English the following conditions:

a) r = radius of a circle (in.) : $2r = 5 + r$

b) W = width of a room (feet): $W + 12 = \frac{3}{2}W$

c) a = age of a student (years): $\frac{3}{4}a = a - 5$

d) A = area of a room (ft^2): $2(A + 14) = 3A$

e) W = width of rectangle (meters), L = length of rectangle (meters): $L - 5 = \frac{1}{2}W$

f) W = width of rectangle (yards), L = length of rectangle (yards): $2L = 4W - 6$

g) V = volume of a pool (m^3): $0 < V \le 2000$

h) A = area of a garden (sq. yards): $A > 150$

i) T = outside temperature (°C): $T \le 0$

For each situation **P1.17** to **P1.33** below:

- Identify and label the variables and main constants including units.

- Draw a diagram (if appropriate) and label it with the variables and constants. (Given diagrams not drawn to scale.)

- Translate requirements into equations and inequalities and write any other additional, related formulas.

P1.17 The Jones family has a pool in the shape of a regular hexagon. They plan to put tile around the edges of the pool. Write a formula for the total amount of tile, in feet, that will go around the pool in terms of the length of a side, S, of the pool.

P1.18   Maria has 304 feet of fence to enclose a rectangular area around her pool. She wants the length of the rectangular area to be four inches less than twice its length.

P1.19   Shawna is creating a banner for a school club. She uses ribbon to outline the design shown below which is made up of equilateral triangles. Write a formula for the total length of ribbon Shawna uses for the outline in terms of the triangle's side length, S.

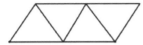

P1.20   A right triangular (artificial) pond is part of a city landscape. The shorter leg is 1/4 times as long as the longer leg. The surface area of the pond is at least 2600 square feet but less than 4000 square feet.

P1.21   Each side of a triangular lot has a different length. The longest side is 4 feet longer than the mid-length side, and the shortest side is 28 inches shorter than the mid-length side. Translate the requirement(s) and write a formula for the perimeter of the triangle.

P1.22   A rectangular swimming area along the lake shore is enclosed by the shore, a rope, and a dock that is 8 meters long. The length of swimming area is 6 meters more than its width and the swimming area must be more than 180 square meters. Translate and write a formula for the length of the rope. (Can you write it in more than one way? Explain.)

P1.23   A rectangular doorway has a temporary support beam across its diagonal. The height of the doorway is 3 times its width. Write a formula for the perimeter of the doorway and a formula for the length of the support beam.

P1.24   Cheryl fenced in a rectangular kennel and used more fence to divide it into three equally sized sections for her dog and his friends as shown. The length, L, of each section is 2 decimeters more than triple its width, W. Write a formula for the total amount of fence used for the kennel.

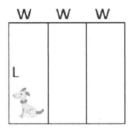

P1.25   The Sunny Days Apartment building has a cylindrical, above ground pool in the yard. The diameter of the pool is six times its height. Translate and write formulas for the circumference of the pool, the area of its circular surface, and volume of the pool.

P1.26   The perimeter of a rectangle is equal to the perimeter of an equilateral triangle. The length of the rectangle is 16 feet more than its width, and a side of the triangle is 14 feet less than twice the width of the rectangle. Translate and write formulas for the perimeters of each figure. What other relation(s) can you express?

P1.27   The top three sales agents in one company are comparing their car sales from last year. Brian sold 160 less than twice the cars that Ariella sold, and Chaya sold 25 fewer cars than Ariella. Translate and write a formula for the total number of cars the three top agents sold.

P1.28 A rectangle and an isosceles triangle have the same perimeter. The length of the rectangle is four times the width. The legs of the triangle are each 21 feet more than the base of the triangle. The base of the triangle is the same as the width of the rectangle.

P1.29 A right triangular pool has a longer leg that is 3 times as long as the shorter leg. The depth of the pool is three fourths of the shorter leg. The volume of the pool is less than 800 cubic feet but more than 500 cubic feet.

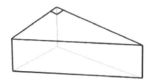

P1.30 Alice, Brian, Colin, and Dave each buy some lottery tickets. All together they buy a total of 208 tickets. Alice buys half as many tickets as Brian. Colin buys three times as many tickets as Brian, and Dave buys four more than four times as many tickets as Brian. The tickets cost $2 each and together they spent less than $500 on the tickets.

P1.31 Sampson bought three items in an electronics store. The most expensive was $15 more than triple the cheapest item. The mid-priced item was $12 less than twice the cheapest item. The total cost of the three items was at least $350.

P1.32 A vending machine has $41.25 in it. There are 255 coins total and the machine only accepts nickels, dimes and quarters. There are twice as many dimes as nickels.

P1.33 In a cash register, there are one-dollar bills, five-dollar bills, ten-dollar bills and twenty-dollar bills. The total amount in the register is $474. There are three times as many ten-dollar bills as twenty-dollar bills. The number of fives is two less than the number of singles. All together, there are 58 bills in the register.

P1.34 I'm thinking of two different positive numbers. Let x = the smaller number and y = the larger number. Translate the following English sentence into an algebraic equation.

"The difference between the two numbers is twenty."

P1.35 Suppose a rectangle has width W and length L. Translate the following sentence into an algebraic equation.

   a) "Five more than four times the width is half of the length."

   b) "The length is three fewer than twice the width."

P1.36 A phone company charges a flat fee of $45 per month plus a 2 cent charge per text message. Write a correct equation for C, the total monthly cost in dollars, in terms of x, the number of text messages.

P1.37 Translate the English expression "four less than a number" into an algebraic expression where n = the number.

Figure 1.8 Deck outline.

1.3 Algebraic Problem Solving with Variables

Launch Exploration

Given a deck outlined in **Fig 1.8** and described below, identify the variables, the constants, and their units (including conversions). Translate the requirements into equations and write any additional formulas related to the situation.

Suppose we are building a rectangular deck. We want the area of the deck to be at least 200 square feet but we use exactly 960 inches of trim for the edges. We also want the length of the deck to be five feet less than twice its width.

Now pick a test value for the width of the rectangle and find the corresponding values for the other variables. Next think of what questions could you ask? For example, what is the area of the rectangle? Make a list of questions.

From Situations to Problems

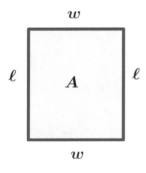

Figure 1.9 $p = \ell + w + \ell + w$

The Deck Problem

Variables: See **Fig 1.9**.

ℓ - the length of the rectangle in feet;

w - the width of the rectangle in feet;

p - the perimeter of the rectangle in feet;

A - area of the rectangle in square feet.

Constants: 960 inches converted to 80 feet.

Equations & Formulas:

The length ℓ of the rectangle is 5 feet less than twice its width ℓ:	$\ell = 2w - 5$
The perimeter of the rectangle is the sum of its side lengths:	$p = 2\ell + 2w$
The area A of a rectangle is the product of its length and its width A:	$A = \ell \times w$
The perimeter of the rectangle is 80 feet:	$p = 80$

Possible questions to answer:

What are the side lengths of the rectangle?

What is the area of the rectangle?

Paint costs $5 per sq. ft; how much would it cost to paint the deck?

Trim costs $4.50 per foot; how much would it cost to trim the deck's outline?

Testing values for w

w	2	7	12
ℓ	−1	9	19
p	2	32	62
A	−2	63	228
p	80	80	80

Table 1.8

A **problem** is a given situation together with one or more *unanswered* questions. **Solving** a problem means answering the questions such that *all* the requirements and constraints in the situation are simultaneously satisfied. Since these requirements and constraints are translated into equations and inequalities, solving a problem is the same thing as answering a question using a *common solution* to a set of equations and inequalities. In this text, we are mostly concerned with equations and will refer to inequalities as *constraints*.

> **Definition 1.8** *System of Equations*
>
> A **system of equations** is a set of equalities grouped together and involving one or more variables. A **common solution** to a system is a set of values for the variables that make *all* the equations in the system true statements. **Solving** a system means finding all of its common solutions.

Going back to the **Launch Exploration**, we have an example of a system of three equations involving three variables:

(**1.3**)
$$\begin{cases} \ell = 2w - 5 \\ p = 2\ell + 2w \\ p = 80 \end{cases}$$

☞ To enclose equations into a system we write a brace { on their left hand side.

Numerical Substitution

When testing the particular value $w = 2$ to find a common solution for the system, we *substitute* this value for w *back* into the system of equations below to get values for the other variables:

$$\begin{cases} \ell = 2w - 5 = 2(2) - 5 = -1 \\ p = 2\ell + 2w = 2(-1) + 2(2) = 2 \end{cases}$$

The values $w = 2$, $\ell = -1$, and $p = 2$ are a common solution to *two* of the three equations in the system.

Given the values $w = 2$, $\ell = -1$, and $p = 2$, if we compare $p = 2$ with the *remaining* equation $p = 80$ we get a *contradiction*. We conclude that these values are *not* a common solution for the system of *three* equations (**1.3**) and thus, cannot be used to answer any of the questions in the problem. Another clue is the fact that some of these values are *negative* and cannot satisfy the constraint of the problem to be a distance since a distance must be positive. Similarly, the other values for w we tested in **Table 1.8** lead to contradictions and this shows that a trial and error method to find solutions is a lottery unless you find a pattern.

Solving Linear Systems by Substitution

Consider the system of equations (**1.3**) for the **Deck Problem** above. We have seen that substituting w by a *numerical* value in the system of equations allows us to find values for the other variables by *evaluations*. This is due in part to the fact that each equation in the system is a formula expressing one variable in terms of others. However, picking an arbitrary value for w most likely leads to a contradiction. So the question is *how do you find the correct value for w?*

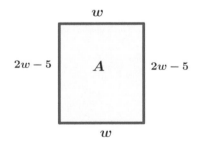

Figure 1.10 $\ell = 2w - 5$

Algebraic Substitution

(Substitution)

To find the correct value for w we keep it as a variable and *substitute* $\ell = 2w - 5$ in $p = 2\ell + 2w$ by replacing ℓ with $2w - 5$.

We get p in terms of w by simplifying the expression *algebraically* (see **Fig 1.10**):[2]

$$p = 2\ell + 2w$$
$$= 2(2w - 5) + 2w$$

(Distribute 2)

$$= 4w - 10 + 2w$$

(Combine like terms)

$$= 6w - 10$$

Now we substitute $p = 6w - 10$ into the last equation $p = 80$. The result is a linear equation involving only the variable w:

(Substitution)

$$6w - 10 = 80.$$

So to find the correct value for w we solve this equation by isolating w:

(Add 10 and divide by 6)

$$6w - 10 = 80, \qquad 6w = 90, \qquad w = 15.$$

(Back substitution)

Substituting $w = 15$ *back* into the system (**1.3**) gives $\ell = 25$, $p = 80$.

The only way to verify whether the set of numerical values $w = 15$, $\ell = 25$, and $p = 80$ satisfy all the requirements and constraints of the problem is to go back to the *original* description and check these values in the context of the situation.

The Verification

The width of the deck is 15 feet.

The length of the deck is 25 feet.

✓ The length of the trim is $25 + 15 + 25 + 15 = 80$ feet.

✓ The length is 25 feet and twice the width is $2(15) = 30$ feet. So the length is $30 - 25 = 5$ feet shorter than twice the width.

✓ All the distances are positive and the area of the deck is $25 \times 15 = 375 > 200$ sq. feet.

The solution has been *validated* and can now be used to answer all the questions we asked. For example, *the dimensions of the deck are 25 ft × 15 ft. The area of the deck is* 375 *square feet. The cost to paint the deck is*

(Related formula)

$$\text{Total cost} = (\text{rate}) \times (\text{area}) = (5) \times (375) = \$1875$$

(Unit analysis)

$$\$ = \left(\frac{\$}{\text{sq. ft}}\right) \times (\text{sq. ft}) = \$$$

(Answer the questions)

Similarly, the cost to trim the deck is $(\text{rate}) \times (\text{perimeter}) = 4.5 \times 80 = 360$ *dollars.*

> **Definition 1.9** *Linear Systems*
>
> An equation or expression is **linear** if the variables and constants are combined by only two operations: 1) addition or 2) multiplication between a variable and a constant. A system is **linear** if all its equations are linear.

Linear equations can be solved by subtractions and divisions; nonlinear equations require different methods. Below we classify each equation into *linear* or *nonlinear* with a short justification:

Linear vs. Nonlinear Equations

Equation	Type	Justification
$5x - 6y = 7$	linear	The multiplications $5x$ and $(-6)y$ and the addition $5x + (-6y)$ give $5x - 6y$.
$t^2 - 3t + 4 = 0$	nonlinear	The square t^2 is a multiplication $t \cdot t$ between two variables.
$\frac{2s-5}{7} = 1 - s$	linear	By distributing the denominator 7 the expression on the left is linear: $\frac{2s-5}{7} = \frac{2}{7}s + \left(-\frac{5}{7}\right)$, and the expression on the right is also linear: $1 - s = 1 + (-1)s$.
$14 - \frac{3}{z} = \frac{1}{z}$	nonlinear	The fraction $\frac{1}{z}$ cannot be written as a constant times the variable z.

☞ Any subtraction can be written as an addition as in $5 - 7 = (5) + (-7)$ and any division can be written as a multiplication as in $-5 \div 7 = (-5) \cdot \left(\frac{1}{7}\right)$.

Our **Deck Problem** involves the linear system (**1.3**) with three equations and three variables and we solved it by using substitutions. It is a mathematical fact that *any* linear system can be solved by *substitutions*.

Linear Systems - More Examples

In each example below we have to solve a problem in which we have to translate the requirements into a linear system of equations. Also the answers to the problem must satisfy certain constraints. As we have seen in the **Deck Problem**, such a *linear problem* can be solved following a strategy summarized in **Table 1.9**.

Solving Linear Problems

Identify the variables, the constants, and their units. Draw a diagram if necessary.

Translate the requirements into equations and identify the type: linear vs. nonlinear.

Solve the linear system of equations by substitutions.

Check the solutions in the context of the situation and answer the questions.

Table 1.9

Example 1.5 Sergio believes he is five years younger than double the age of Chloe, and Chloe believes she is five years older than half the age of Sergio. Are they both right?

Solution: Let S be Sergio's age in years. Let C be Chloe's age in years.   (Variables)

$$\begin{cases} S = 2C - 5 \\ C = \frac{1}{2}S + 5 \end{cases}$$ (Linear system)

Substitute $S = 2C - 5$ into $C = \frac{1}{2}S + 5$ and get

$$C = \tfrac{1}{2}S + 5$$
$$= \tfrac{1}{2}(2C - 5) + 5$$
$$= \tfrac{1}{2} \cdot 2C - \tfrac{1}{2} \cdot 5 + 5$$
$$= C + \frac{5}{2}.$$

(Substitute)

(Distribute $\frac{1}{2}$)

(Combine constant terms)

Solve the linear equation $C = C + \frac{5}{2}$ by subtracting C from both sides and get a *contradiction* $0 = \frac{5}{2}$.

(Solve a linear equation)

Since the system has NO solutions, we conclude that *Sergio and Chloe cannot both be right.*

(Answer the question)

Example 1.6 Abe, Bella, and Carlos have all earned some college credits. Bella has earned 17 credits more than Abe, and Carlos has earned 9 less than triple the number of Abe's credits. Altogether, they have earned a total of 183 credits. How many credits has each earned?

Solution: Let A be the number of credits that Abe has earned. Let B be the number of credits that Bella has earned. Let C be the number of credits that Carlos has earned. The main constant is 180 total number of credits.

(Variables and constants)

$$\begin{cases} B = A \;+\; 17 \\ C = 3A \;-\; 9 \\ 183 = A + B + C \end{cases}$$

(Linear system)

Substitute $B = A + 17$ and $C = 3A - 9$ into $183 = A + B + C$ and get[3]

$$183 = A + B + C$$
$$= A + (A + 17) + (3A - 9)$$
$$= A + A + 17 + 3A - 9$$
$$= 5A + 8.$$

(Substitute)

(Combine like terms)

Solve the linear equation $183 = 5A + 8$ by subtracting 8 and dividing by 5:

$$183 = 5A + 8, \qquad 175 = 5A, \qquad A = 175/5 = 35.$$

(Solve a linear equation)

Substitute $A = 35$ back into the system and find the values of the other variables:

$$\begin{cases} A = 35 \\ B = A + 17 = (35) + 17 = 52 \\ C = 3A - 9 = 3(35) - 9 = 96 \end{cases}$$

(Back substitution)

Abe has earned 35 credits.

Bella has earned 52 credits.

Carlos has earned 96 credits. (Answer the question)

Bella has earned $52 - 35 = 17$ credits more than Abe. ✓

Triple number of Abe's credits is $3(35) = 105$ and Carlos earned $105 - 96 = 9$ credits less than that. ✓

Altogether, they earned $35 + 52 + 96 = 183$ credits. ✓

Example 1.7 Mateo has 14 meters of fence for a rectangular pen alongside of his house. He only needs to fence three sides such that the length is 1 meter less than triple the width. The pen is for his dog who needs more than 20 square meters to be happy. What are the dimensions of the pen?

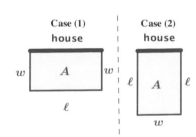

Figure 1.11 Which one?

Solution: Let ℓ be the length of the pen in meters, w the width of the pen in meters, f the length of the fence in meters; A the area of the pen in square meters. The main constants are 14 meters of fence and 20 square meters of area. We have two cases (see **Fig 1.11**):

$$\textbf{Case (1)} \begin{cases} f = 14 \\ \ell = 3w - 1 \\ f = \ell + 2w \\ A = \ell \times w \end{cases} \quad \text{OR} \quad \textbf{Case (2)} \begin{cases} f = 14 \\ \ell = 3w - 1 \\ f = 2\ell + w \\ A = \ell \times w \end{cases}$$

Case (1). *Mateo has 14 meters of fence for a rectangular pen whose length is alongside of his house. He only needs to fence three sides such that the length is 1 meter less than triple the width. The pen is for his dog who needs more than 20 square meters to be happy. What are the dimensions of the pen?*

Substitute $f = 14$ and $\ell = 3w - 1$ into $f = \ell + 2w$ and get

$$f = \ell + 2w$$
$$14 = (3w - 1) + 2w \qquad \text{(Substitution)}$$
$$14 = 5w - 1. \qquad \text{(Combine like terms)}$$

Solve the linear equation $14 = 5w - 1$ by adding 1 and dividing by 5:

$$14 = 5w - 1, \qquad 15 = 5w, \qquad w = 15/5 = 3. \qquad \text{(Solve a linear equation)}$$

Substitute $w = 3$ back into the system and find the other variables:

$$\textbf{Case (1)} \begin{cases} w = 3 \\ \ell = 3w - 1 = 3(3) - 1 = 8 \\ f = \ell + 2w = 8 + 2(3) = 14 \\ A = \ell \times w = 8 \times 3 = 24 \end{cases}$$

(Back substitution)

The length of the pen is 8 meters.

The width of the pen is 3 meters.

✓ The length of the fence is $3 + 8 + 3 = 14$ meters.

The area of the pen is $8 \times 3 = 24$ square meters and is *at least* 20 square

✓ meters.

(Answer the question)   *The dimensions of the pen that satisfies all the constraints are 8 m × 3 m.*

Case (2). *Mateo has 14 meters of fence for a rectangular pen whose width is alongside of his house. He only needs to fence three sides such that the length is 1 meter less than triple the width. The pen is for his dog who needs more than 20 square meters to be happy. What are the dimensions of the pen?*

Substitute $f = 14$ and $\ell = 3w - 1$ into $f = 2\ell + w$ and get

(Substitution)

$$f = 2\ell + w$$
$$14 = 2(3w - 1) + w$$
$$14 = 6w - 2 + w$$

(Combine like terms)

$$14 = 7w - 2.$$

Solve the linear equation $14 = 7w - 2$ by adding 2 and dividing by 7:

(Solve a linear equation)

$$14 = 7w - 2, \qquad 16 = 7w, \qquad w = \frac{16}{7} = 2\frac{2}{7}.$$

Back substitute $w = 16/7$ into the system and find the other variables:

Case (2)
$$\begin{cases} w = \frac{16}{7} \\ \ell = 3w - 1 = 3 \cdot \left(\frac{16}{7}\right) - 1 = \frac{48}{7} - \frac{7}{7} = \frac{41}{7} = 5\frac{6}{7} \\ f = 2\ell + w = 2 \cdot \left(\frac{41}{7}\right) + \left(\frac{16}{7}\right) = \frac{82}{7} + \frac{16}{7} = \frac{98}{7} = 14 \\ A = \ell \times w = \left(\frac{41}{7}\right) \times \left(\frac{16}{7}\right) = \frac{656}{49} = 13\frac{19}{49} \end{cases}$$

(Back substitution)

✗

(Answer the question)   Since the area of the pen is $13\frac{19}{49}$ square meters and thus, it does NOT satisfy the constraint to be at least 20 square meters, *in this case we do NOT have a solution to the problem.*

Power Equation - A Basic Example of a Nonlinear Equation

A **power equation** is an equation of the form $x^n = \pm a$ where x is a variable, $n > 1$ is a natural number, and a is a *positive* real number. The equation $x^n = a$ has a *positive* real root denoted by $x = a^{1/n} = \sqrt[n]{a}$ and called the **principal n-th root of a** or the **radical of index n of a**. If the exponent n is *odd*, this is the *only* real number solution of the equation $x^n = a$. If the exponent n is *even*, the equation $x^n = a$ has exactly *two* real number solutions. The second one is the **negative n-th root of a**, $x = -a^{1/n} = -\sqrt[n]{a}$ which is of course a *negative* number.

☞ The *n-th power* of a quantity q is written q^n and means n copies of q multiplied together. For example,

$$q^2 = qq, \quad q^3 = qqq, \quad \text{etc.}$$

Square Power Equations - Examples

a) The equation $x^2 = 49$ has exactly two real solutions: the *positive square root*

$$x = 49^{1/2} = \sqrt{49} = 7$$

and the *negative square root*

$$x = -(49)^{1/2} = -\sqrt{49} = -7.$$

Indeed, if $x = -7$ or $x = 7$, then we have

$$x^2 = (-7)^2 = (-7)(-7) = 49 \qquad \text{or} \qquad x^2 = (7)^2 = 7 \cdot 7 = 49.$$

b) The equation $x^2 = 3$ has two real solutions $x = \sqrt{3}$ and $x = -\sqrt{3}$ which are not integers. To check that $x = -\sqrt{3}$ is a solution we take its square:

$$x^2 = \left(-\sqrt{3}\right)^2 = \left(-\sqrt{3}\right)\left(-\sqrt{3}\right) = \sqrt{3} \cdot \sqrt{3} = \left(\sqrt{3}\right)^2 = 3.$$

c) The equation $x^2 + 25 = 0$ has no real solutions since if you write it as $x^2 = -25$, the square of any real number cannot equal a negative number.

☞ $x^{\text{even}} = $ positive has two real solutions; $x^{\text{even}} = $ negative has no real solutions.

The equation $x^n = -a$ has *no* real number solution if the exponent n is *even* since an even power of a real number is always *nonnegative* and thus, it cannot be equal to the negative number $-a$. However, if the exponent n is *odd*, then the equation $x^n = -a$ has exactly *one* real number solution, the n-th root of a, $x = -a^{1/n} = -\sqrt[n]{a}$.

☞ $\sqrt[n]{-11} = -\sqrt[n]{11}$ is a real number if n is *odd* as in $\sqrt[3]{-11} = -\sqrt[3]{11}$, but $\sqrt[n]{-11}$ is not a real number if n is *even*; $\sqrt{-11} \neq -\sqrt{11}$.

Cubic Power Equations - Examples

a) The equation $x^3 = 27$ has only *one* real solution, the cube root $x = \sqrt[3]{27} = 3$ since $3^3 = (3)(3)(3) = 27$.

b) The equation $x^3 = -125$ has a *unique* real number solution, the *cube root* $x = -(125)^{1/3} = -\sqrt[3]{125} = -5$. To check it we take the cube of $x = -5$ and get

$$x^3 = (-5)^3 = (-5)(-5)(-5) = -5 \cdot 5 \cdot 5 = -125.$$

c) The equation $x^3 = -11$ has only the real solution $x = -\sqrt[3]{11}$, which is not an integer:

$$x^3 = \left(-\sqrt[3]{11}\right)^3 = \left(-\sqrt[3]{11}\right)\left(-\sqrt[3]{11}\right)\left(-\sqrt[3]{11}\right) = -\sqrt[3]{11} \cdot \sqrt[3]{11} \cdot \sqrt[3]{11} = -\left(\sqrt[3]{11}\right)^3 = -11.$$

☞ $x^{\text{odd}} = $ positive has one real solution; $x^{\text{odd}} = $ negative has one real solution.

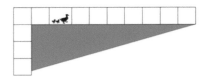

Figure 1.12 Ducks by a pond

(Linear equations)

(Area formula - nonlinear)

(Multiply by 8)

(Solve a power equation)

(Answer the question)

✓

✓

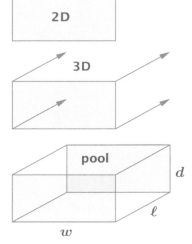

Figure 1.13 Building a Pool

Example 1.8 A right triangular (artificial) pond is part of a city landscape. See **Fig 1.12**. The shorter leg is 1/4 times as long as the longer leg. How far will a duck walk from one end of the longer leg to the other if the area of the pond is 3,000 sq. feet?

Solution: Let s be the length of the shorter leg in feet, ℓ the length of the longer leg in feet, and A the area of the pond in sq. feet.

The translated requirements and related formulas are

$$\begin{cases} s = \frac{1}{4}\ell \\ A = 3000 \\ A = \frac{1}{2}s\ell \end{cases}$$

Substitute $s = \frac{1}{4}\ell$ and $A = 3000$ into the equation $A = \frac{1}{2}s\ell$ and get

$$3000 = \frac{1}{2}\left(\frac{1}{4}\ell\right)\ell = \frac{1}{8}\ell^2 \qquad \text{or} \qquad 24000 = \ell^2.$$

Now solve the equation $\ell^2 = 24000$ for ℓ by taking the square roots

$$\ell^2 = 24000, \qquad \ell = \sqrt{24000}, \qquad \cancel{\ell = -\sqrt{24,000}.}$$

A duck will walk $\sqrt{24000} \approx 155$ *feet* since a distance is positive.

The shorter leg is $\frac{1}{4}\sqrt{24000} \approx 38.7$ feet.

The area of the pond is half the product of the legs or

$$\frac{1}{2}\left(\frac{1}{4}\sqrt{24000}\right)\left(\sqrt{24000}\right) = \left(\frac{1}{8}\right)\left(\sqrt{24000}\right)^2 = \frac{24000}{8} = 3000 \text{ sq. feet.}$$

Example 1.9 Pam has a box-shaped lap pool two times wider than it is deep and six times longer than it is wide. The water charge to fill the pool is $6 per 100 cubic feet and she wants to pay no more than $100.

a) If Pam spends $92.16, how deep is the pool?

b) Is a 5 feet deep pool affordable for Pam?

We treat each question as a separate problem. See **Fig 1.13**.

a) *Pam has a box-shaped lap pool two times wider than it is deep and six times longer than it is wide. The water charge to fill the pool is $6 per 100 cubic feet. If Pam spends $92.16, how deep is the pool?*

Solution: Let w be the width, ℓ the length, and d the depth of the pool in feet. Let V be the volume of the pool in cubic feet and C the total cost to fill in the pool in dollars. The main constants are $6 per 100 cubic feet (rate) and $92.16 total cost. (Variables and constants)

$$\begin{cases} w = 2d \\ \ell = 6w \\ V = w\ell d \\ C = \frac{6}{100}V \end{cases}$$

(Linear equation)
(Linear equation)
(Volume formula - nonlinear)
(Linear equation)

Substitute $w = 2d$ in $\ell = 6w$ and get the formula

$$\ell = 6w = 6(2d) = 12d.$$ (Substitute and simplify)

Substitute $w = 2d$ and $\ell = 12d$ in $V = w\ell d$ and get the formula

$$V = w\ell d = (2d)(12d)d = 24d^3.$$ (Substitute and simplify)

Substitute $V = 24d^3$ into $C = \frac{6}{100}V$ and get the formula

$$C = \frac{6}{100}V = \frac{6}{100}\left(24d^3\right) = 1.44d^3.$$

To find the depth of the pool when the cost is $92.16, substitute $C = 92.16$ into the equation $C = 1.44d^3$, divide by 1.44, and solve for d:

$$92.16 = 1.44d^3, \qquad d^3 = 92.16/1.44 = 64, \qquad d = \sqrt[3]{64} = 4.$$ (Solve a power equation)

Now we look back to the *original* description and verify all the requirements.

If the depth of the pool is 4 feet, the width is 8 feet, twice the depth. ✓

The length is 6 × 8 = 48 feet, six times the width. ✓

The volume is 4 × 8 × 48 = 1536 cubic feet. (Volume formula)

The cost to fill in the pool is

$$\text{Total cost} = (\text{rate}) \times (\text{volume}) = \left(\tfrac{6}{100}\right) \times (1536) = \$92.16$$ ✓

$$\$ = \left(\frac{\$}{\text{cu. ft}}\right) \times (\text{cu. ft}) = \$$$ (Unit analysis)

If the cost is $92.16, then *the depth of the pool is 4 feet*. (Answer the question)

b) *Pam has a box-shaped lap pool two times wider than it is deep and six times longer than it is wide. The water charge to fill the pool is $6 per 100 cubic feet and she wants to pay no more than $100. Is a 5 ft deep pool affordable?*

If the depth of the pool is 5 feet, the width is 10 feet, twice its depth. ✓

The length is 60 feet, six times the width. ✓

The volume is 5 × 10 × 60 = 3000 cubic feet. (Volume formula)

The total cost is $\frac{6}{100} \times 3000 = 180$ dollars. (Rate × volume)

Since the cost is $180 and 180 > 100, *this pool is not affordable*. (Answer the question)

Wrapping Up

After we translate situations with constraints into systems of equations, we show how to reduce these systems to a linear or a power equation in one variable by substitutions. After we solve for that variable, we substitute its numerical value back into the system and find the values of the other variables. In the end we verify all the constraints and answer the questions. We summarize this problem solving strategy as follows

- *Understand the problem:* Identify the variables, the constants, and their units. Draw a diagram if necessary.

- *Devise a plan:* Translate the requirements into equations and identify the type: linear vs. nonlinear.

- *Carry out the plan:* Solve the system of equations by substitutions.

- *Look back:* Check the solution(s) in the context of the situation and answer the questions.

Exercises

Exercises for 1.3 Algebraic Problem Solving with Variables

P1.38 Solve for the unknowns.

  a) r = radius of a circle (in.) : $2r = 5 + r$

  b) W = width of a room (feet): $W + 12 = \frac{3}{2}W$

  c) a = age of a student (years): $\frac{3}{4}a = a - 5$

  d) A = area of a room (ft^2): $2(A + 14) = 3A$

For each situation **P1.39** to **P1.55** below:

- Identify and label the variables and main constants including units. Draw a diagram if appropriate.

- Translate requirements into equations and/or inequalities and write any other additional, related formulas.

- Identify equations by type: linear/nonlinear. Solve the equation or the linear system of equations by substitutions.

- Check the solutions in the context of the situation and answer the questions.

P1.39 The Jones family has a pool in the shape of a regular hexagon. If it takes $49\frac{1}{2}$ feet of tile to go around the edge of the pool, find the length, S, of a side of the pool.

P1.40 Shawna is creating a banner for a school club. She uses ribbon to outline the design shown below which is made up of equilateral triangles. The bottom edge of the banner is 50 inches long. How much total ribbon is used to outline the banner?

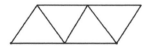

P1.41 Maria has 304 feet of fence to enclose a rectangular area around her pool. She wants the length of the rectangular area to be four inches less than twice its length. If the length is 5 inches, what is the width?

P1.42 Each side of a triangular lot has a different length. The longest side is 4 feet longer than the mid-length side, and the shortest side is 28 inches shorter than the mid-length side. The fence around the lot is 100 feet. How long is each side of the lot?

P1.43 A rectangular swimming area along the lake shore is enclosed by the shore, a rope, and a dock that is 8 meters long. The length of swimming area is 6 meters more than its width and the swimming area must be more than 180 square meters. If the rope is 34 meters long, what are the dimensions of the swimming area?

P1.44 A rectangular doorway has a temporary support beam across its diagonal. The height of the doorway is 3 times its width. If the beam across the diagonal is 10 feet long, how wide is the door?

P1.45 Cheryl fenced in a rectangular kennel and used more fence to divide it into three equally sized sections for her dog and his friends as shown. The length, L, of each section is 2 decimeters more than triple its width, W. If the total amount of fence used for the kennel is 45.8 meters, what is the area of each section?

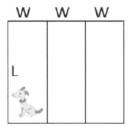

P1.46   The Sunny Days Apartment building has a cylindrical, above ground pool in the yard. The diameter of the pool is six times its height, and the volume is 2,576.5 cubic feet. How deep is the pool?

P1.47   The perimeter of a rectangle is equal to the perimeter of an equilateral triangle. The length of the rectangle is 16 feet more than its width, and a side of the triangle is 14 feet less than twice the width of the rectangle. What are the dimensions of each shape?

P1.48   The top three sales agents in one company are comparing their car sales from last year. Brian sold 160 less than twice the cars that Ariella sold, and Chaya sold 25 fewer cars than Ariella. If the three of them sold 515 cars altogether, how many did each sell?

P1.49   A rectangle and an isosceles triangle have the same perimeter. The length of the rectangle is four times the width. The legs of the triangle are each 21 feet more than the base of the triangle. The base of the triangle is the same as the width of the rectangle. What are all of the side lengths for each figure?

P1.50   A right triangular pool has a longer leg that is 3 times as long as the shorter leg. The depth of the pool is three fourths of the shorter leg. If the volume of the pool is 576 cubic feet then how deep is the pool?

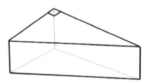

P1.51   Alice, Brian, Colin, and Dave each buy some lottery tickets. Alice buys half as many tickets as Brian. Colin buys three times as many tickets as Brian, and Dave buys four more than four times as many tickets as Brian. If the tickets cost $2 each and together they spent exactly $416, then how many tickets did they each buy?

P1.52   Sampson bought three items in an electronics store. The most expensive was $15 more than triple the cheapest item. The mid-priced item was $12 less than twice the cheapest item. If the total cost of the three items was $387, what was the cost of each of the three items?

P1.53   A vending machine has $41.25 in it. There are 255 coins total and the machine only accepts nickels, dimes and quarters. There are twice as many dimes as nickels. How many of each coin is in the machine?

P1.54   Ms. Smith is building a circular pool right next to a rectangular patio as shown in the figure. The width of the rectangular patio is exactly the same as the diameter of the pool, and the length of the patio is three times its width. It took 44 feet of tile to go around the edge of the pool. How many square feet of tile will it take to cover the patio?

P1.55   In a cash register, there are one-dollar bills, five-dollar bills, ten-dollar bills and twenty-dollar bills. The total amount in the register is $474. There are three times as many ten-dollar bills as twenty-dollar bills. The number of fives is two less than the number of singles. All together, there are 58 bills in the register. How many of each bill is there?

P1.56   Solve the linear equations for the unknown.

   a) $-3(y+2)-20=-5$

   b) $4-2(t-5)=3t-4(t+2)$

   c) $x-20+4x+3x=-2+7x-11$

   d) $\frac{9}{2}x+\frac{2}{3}=\frac{5}{3}+\frac{3x}{5}$

   e) $\frac{1}{2}(2x+5)=\frac{1}{4}(8x-1)$

   f) $\frac{3t-8}{2}=\frac{3}{2}-\frac{t}{3}$

   g) $5x+2-x=10(x-1)-2(3x-6)$

   h) $0(23x-5)=11(8x-8x)$

   i) $6p+4(2p+3-p)=2(6+5p)+7$

   j) $24a-22=-4(1-6a)$

P1.57 Solve the power equations for the unknown:

　　a) $x^3 = 64$

　　b) $z^3 = -64$

　　c) $t^3 + 16 = 0$

　　d) $x^2 = -16$

　　e) $v^2 = 81$

　　f) $s^2 - 2 = 0$

　　g) $4r^2 - 9 = 0$

　　h) $4r^2 + 9 = 0$

　　i) $27S^3 = 125$

　　j) $17L^3 - 34 = 0$

P1.58 [Harvard 1869] A man bought a watch, a chain, and a medallion with $216. The watch and the medallion together cost three times the chain. The chain and the medallion together cost half the watch. What was the price of each?

P1.59 One of the club teams on campus is making a flag for their team as shown in the figure below (not drawn to scale.) The blue section is a square. The striped sections are all exactly the same size, and the length of each striped section is six times as long as the width of each striped section. The club uses black ribbon to trim around the outside edge and to trim all of the lines between sections. If they use 351 inches of black ribbon, what is the area for the entire flag?

P1.60 Nat used 366 feet of fencing to enclose a rectangular garden and to put a divider between the two halves of the garden - the vegetable side and the herb side. The width of the garden is 18 feet less than its total length (diagram not drawn to scale). Figure out the dimensions of the entire garden and then find the area of each half of the garden.

P1.61 Calculate the area of each figure below.

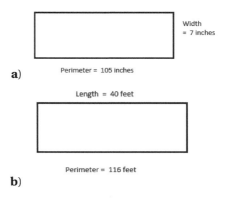

Width = 7 inches

Perimeter = 105 inches

a)

Length = 40 feet

Perimeter = 116 feet

b)

P1.62 Calculate the perimeter of each figure below.

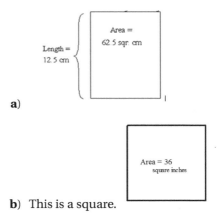

Length = 12.5 cm

Area = 62.5 sqr. cm

a)

Area = 36 square inches

b) This is a square.

P1.63 Calculate the area of the figure below.

Circumference = 18π inches

1.4   Relationships between Variables - Introduction to Functions

In the first sections of this book we analyzed situations and the relationships among their many variables and constants. For the remainder of this course we will focus on *only two* variables in a given situation and the relationship between them. We pair related variables all the time - first names with last names, days of the week with classes, hours worked in a week with gross weekly pay, etc. Any matching between two variables is called a *relation* where one of the variables is considered the input of the relation and the other the output.

Launch Exploration

A club secretary is organizing information about club members. She matches each member with three variables as shown in the tables below:

Input Student	Output Birth Month	Input Student	Output ID Number	Input Student	Output Subject
Om	Apr	Om	8310	Om	MATH
Matt	May	Matt	5108	Matt	CHEM
Lubna	Jun	Lubna	4195	Lubna	HIST
Asya	Apr	Asya	0981	Om	ENG
Cher	Dec	Cher	4033	Asya	ART
Drake	Oct	Drake	7206	Cher	PSY
				Drake	ENG

Table 1.10 Analyzing relations

Each table displays a list of ordered pairs of the form (**input, output**). Any set of ordered pairs is a **relation** between two variables. So each table made by the club secretary represents a relation. Identify and label the two variables in each relation. The set of inputs of a relation is called the **domain** of the relation and the set of outputs is called the **range**. Determine the domain and range of each relation. Now we define a special kind of relation:

A relation is a **function** *if each input is matched with one and only one output.*

Using this definition, determine which of the three tables above represent functions and which do not. Explain your reasoning.

From Relations to Functions

An **ordered pair** is a list of two items, a first and a second, written as (**first, second**) where the order matters. For example, in an ordered pair of the form (First Name, Last Name) we have (Carter, Joseph) ≠ (Joseph, Carter) since the two ordered pairs represent two people with different last names. Relations are sets of ordered pairs. The domain and range of a relation are also sets. A **set** is a collection of objects called **elements**. The elements of a set could be names, numbers, months, subjects, ordered pairs, etc.

Definition 1.10 *Relations and their Domain and Range*

A **relation** is any set of ordered pairs of the form (**input, output**), which can be displayed in an **input-output table**. The **domain** of a relation is the set of all allowable inputs and the **range** is the set of all outputs.

One way to express a set is to assign it a name and list its elements separated by commas inside braces, without repetitions. For example, in the **Launch Exploration** we name the student - birth Month relation by the letter $\mathcal{M}$; student - ID Number by $\mathcal{N}$; and student - Subject by $\mathcal{S}$. The sets are

$\mathcal{M}$ = {(Om, Apr), (Matt, May), (Lubna, Jun), (Asya, Apr), (Cher, Dec), (Drake, Oct)}

Domain of $\mathcal{M}$ = { Om, Matt, Lubna, Asya, Cher, Drake }
Range of $\mathcal{M}$ = { Apr, May, Jun, Dec, Oct }

$\mathcal{N}$ = {(Om, 8310), (Matt, 5108), (Lubna, 4195), (Asya, 0981), (Cher, 4033), (Drake, 7206)}

Domain of $\mathcal{N}$ = { Om, Matt, Lubna, Asya, Cher, Drake }
Range of $\mathcal{N}$ = { 8310, 5108, 4195, 0981, 4033, 7206 }

$\mathcal{S}$ = {(Om, MATH), (Matt, CHEM), (Lubna, HIST), (Om, ENG), (Asya, ART), (Cher, PSY), (Drake, ENG)}

Domain of $\mathcal{S}$ = { Om, Matt, Lubna, Asya, Cher, Drake }
Range of $\mathcal{S}$ = { MATH, CHEM, HIST, ENG, ART, PSY }

Definition 1.11 *Functions as Sets of Ordered Pairs*

A **function** is a set of ordered pairs (relation) of the form (**input, output**) such that each allowable input is paired with *one and only one* output.

In the **Launch Exploration** examples, the input variable in all three relations is the student club member which we label by the letter x. For the relation $\mathcal{M}$ the output variable is m - birth month. For the relation $\mathcal{N}$ the output variable is n - ID number. For the relation $\mathcal{S}$ the output variable is s - subject.

In the relation $\mathcal{M}$, two students, Om and Asya, have the same birth month, April, which means that we have two inputs with the same output. This is guaranteed to happen for any club with more than 12 members. Does this mean that the relation $\mathcal{M}$ is not a function? No. It is okay for inputs of a function to share the

Relation ($\mathcal{M}$)	
Input Student x	**Output Birth Month** m
Om	Apr
Matt	May
Lubna	Jun
Asya	Apr
Cher	Dec
Drake	Oct

Table 1.11 Function

Relation ($\mathcal{N}$)	
Input Student x	**Output ID Number** n
Om	8310
Matt	5108
Lubna	4195
Asya	0981
Cher	4033
Drake	7206

Table 1.12 Function

Relation ($\mathscr{S}$)	
Input Student x	Output Subject s
Om	MATH
Matt	CHEM
Lubna	HIST
Om	ENG
Asya	ART
Cher	PSY
Drake	ENG

Table 1.13 NON - Function

☞ **Function Notations**

OK:
$y = y(x)$ or $y = f(x)$ or
$d = h(t)$

NOT OK:
$y = yx$ or $y = fx$ or $d = ht$

DON'T GO THERE:
$y(x) = y \cdot x$ or $f(x) = f \cdot x$ or
$h(t) = h \cdot t$

Function notation should not be confused with multiplication.

same output. Because each student can have only one birth month, each input has only one output, so $\mathscr{M}$ IS a function. See **Table 1.11**.

In the relation $\mathscr{N}$, each student has a unique ID number in order to be identified. In this case, each input has one and only one output. Therefore $\mathscr{N}$ is a function. Since two students cannot have the same ID number, two inputs cannot share the same output. A function with this property is called a **one-to-one** function. For example, the function $\mathscr{M}$ is NOT one-to-one. See **Table 1.12**.

In the relation $\mathscr{S}$, there is a student, Om, who is learning TWO subjects, MATH and ENG, which means $\mathscr{S}$ has an input with more than one output. Therefore $\mathscr{S}$ is not a function. See **Table 1.13**.

Function Notation - How to Write about Functions

A function can be thought of as *a rule which assigns to each input in its domain, one and only one output in its range.* If the input variable is labeled by x, the output variable by y, and the rule by f, then the output variable can also be referred to as $y = f(x)$. This new label is called **function notation** and has the advantage that it shows the input which matches the output:

(**1.4**) $f(\text{input}) = \textbf{output.}$

This is read, "f of the input equals the output." For example, in **Table 1.11**, the function rule labeled $\mathscr{M}$ matches club members with their birth months. The input is labeled by x, and the output by m. Suppose the club secretary is planning a birthday celebration for the month of Dec. We know the output m = Dec for our birth month function, but when we only write m = Dec, we have no way of knowing which club member has a birthday in Dec. However, if we use the function notation

$$\mathscr{M}(x) = m$$

then for x = Cher and m = Dec, we can write

$$\mathscr{M}(\textbf{Cher}) = \textbf{Dec} ,$$

which tells us the output value is Dec when the input value is Cher. This gives the ordered pair (Cher, Dec). Having the entire ordered pair is essential when we analyze real-world functions and so we will continue to use function notation throughout this book.

Notice that $\mathscr{M}(\text{Mario})$ is not defined since Mario is not in the domain of the function $\mathscr{M}$. Can you find the input x for which $\mathscr{M}(x)$ = Apr? The answer is x = Om or x = Asya. What about $\mathscr{M}(x)$ = Om? In this case there is no such input, since Om is NOT in the range of the function $\mathscr{M}$.

From Situations to Functions

In many situations with two related variables, the values of one variable depend on the values of the other. For example, weekly pay depends on hours worked, total cost of candies depends on the number of candies purchased, the amount of water left in a pool depends on draining time, etc. In a function, the outputs depend on inputs and so we call the output variable the *dependent variable* while the input variable is the *independent variable*.

> ### Definition 1.12 *Independent vs. Dependent Variables*
>
> In a function, the **independent variable (IV)** is the input variable, which is independently changing. The **dependent variable (DV)** is the output variable, which is affected by the change in the independent variable.

Preliminary Function Analysis

Identify and label the independent and the dependent variables.

Determine whether the variables are continuous or discrete.

Start a table to list some ordered pairs of the form (input, output).

Determine and write the domain and range.

Table 1.14

Example 1.10

Tyler earns $12 per hour at his part-time job and he works between 0 and 30 hours each week. We assume that Tyler is payed in one hour increments only.

Do the tasks in the margin **Table 1.14** and then answer the following questions:

- What is Tyler's weekly pay if he works 18 hours?
- How many hours does Tyler work to get paid $120?

Express your answers using both ordered pairs and function notation and enter them in the table.

Solution: The weekly pay depends on the hours worked. So we choose the independent and dependent variables to be

$$\text{Independent Variable (Input):} \quad t = \textbf{hours worked}$$
$$\text{Dependent Variable (Output):} \quad W(t) = \textbf{weekly pay in \$}$$

If we round the values for t to the nearest integer for example, then we consider t to be **discrete.** So, the real world domain of the function is the set of all allowable inputs specified in the situation from 0 to 30 hours:

$$\textbf{Domain} = \{\, 0\,\text{hrs}\,,\, 1\,\text{hr},\, 2\,\text{hrs},\, 3\,\text{hrs},\, ...,\, 30\,\text{hrs}\,\}$$

The weekly pay $W = W(t)$ is also **discrete** as the outputs are given by the formula:

$$\textbf{Weekly pay} = (\,\textbf{rate per hour}\,) \times (\,\textbf{hours worked}\,)$$

So, the real world range of the function is the set of all outputs:

$$\textbf{Range} = \{\, \$0,\, \$12,\, \$24,\, \$36,\, ...,\, \$360\,\}$$

Tyler's weekly pay if he works 18 hours is $W = (\textbf{\$12/hr}) \times (\,18\,\text{hrs}\,) = \$216$. As W is also the name of the function, in function notation we write $W(\,18\,\text{hrs}) = \$216$ and as an ordered pair, ($18\,\text{hrs}$, $\$216$).

For a weekly pay of $120, Tyler works $t = (\,\$120\,) \div (\textbf{\$12/hr}) = 10\,\text{hrs}$. In function notation, we write $W(\,10\,\text{hrs}) = \$120$ and as an ordered pair, ($10\,\text{hrs}$, $\$120$). See **Table 1.15** for a sample of ordered pairs, including the answers to the questions.

IV (In) t (hrs)	DV (Out) $W(t)$ ($)	Pair $(t, W(t))$
0	0	(0,0)
1	12	(1,12)
2	24	(2,24)
3	36	(3,36)
10	120	(10,120)
18	216	(18,216)
30	360	(30,360)

Table 1.15 Data Sample.

How to Visualize Functions - Graphs

To visualize a relation (function) we can create a *graph* by plotting its ordered pairs in the plane. To do that, we represent the inputs on an *horizontal axis* labeled by the independent variable and the outputs on a *vertical axis* labeled by the dependent variable. An ordered pair (*input, output*) is represented by a point in a plane whose *horizontal coordinate* is the input and the *vertical coordinate* is the output. See **Fig 1.14**.

When the variables take numerical values each axis is a number line. In this case, recall that the positive direction on the horizontal axis is to the right and on the vertical axis is upwards. The two axes meet at the common zero point called the *origin* of the axes.

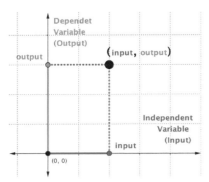

Figure 1.14 Plot a point in a plane.

Definition 1.13 *The Graph of a Relation & Scatter Plots*

The **graph** of a relation is the set of ALL points in a plane with **coordinates** ordered pairs **(input, output)** of the relation. A finite sample of such points is called a **scatter plot** and the pairs are **(data) points** of the relation.

Going back to **Example 1.10**, the horizontal axis is labeled by the independent variable t, the hours worked, and the vertical axis is labeled by the dependent variable W, the weekly pay. A scatter plot for this function is obtained by plotting the ordered pairs from **Table 1.15** in a plane:

☞ **Scaling axes:**

Each axis has its own scale chosen to fit the data.

For example, to fit a range from $0 to $360 with six tick marks we choose a scale of $360 \div 6 = $60 per tick mark.

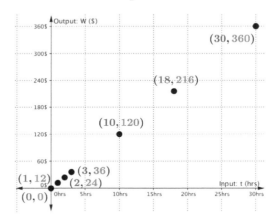

Figure 1.15 Tyler's weekly pay

The horizontal axis is scaled using tick marks every 5 hours, and the vertical axis is scaled using tick marks every 60 dollars. Both the table and the graph show a sample of 7 data points. Since the variables are discrete, the full graph will have 31 isolated dots corresponding to the 31 data points of the entire function.

Recognizing Functions from Tables and Graphs

Given an input-output table of a relation, to decide whether the relation is a function or not based on that table, we need to check whether each input has exactly one output. To do that, we search for inputs which are repeated in the first column and check whether their outputs are the same or not. If we find at least one input with two *different* outputs, the relation is not a function. If such an example cannot be found, we conclude that the relation is a function.

Hours Candle Burns	Height of Candle (cm)
1	28
4	26
7	24
10	22
13	20
16	18

(a)

x	Square Roots of x
0	0
1	1
1	−1
4	2
4	−2
9	3
9	−3

(b)

Number of Toppings	Total Cost of Toppings ($)
4	0
6	3
8	6
10	9
12	12

(c)

Figure 1.16 Is it a Function?

Example 1.11 For each table above plot a graph. Label axes and scales. Determine whether or not it represents a function and EXPLAIN why.

IF it IS a function:

Identify and label the independent and dependent variables and give the function a name.

Then express at least two of the ordered pairs in function notation.

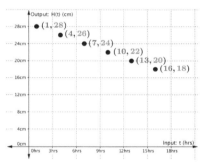

Figure 1.17 Candle **(a)**.

Solution: The graphs are displayed in the margin. The relation **(b)** **is not a function** since the ordered pairs (1, 1) and (1, −1) have the same input: 1, but two outputs: 1 and −1. The relations **(a)** and **(c)** **are functions** since each input has exactly one output.

For the function **(a)** the independent and dependent variables are

$$\text{IV (Input):} \quad t = \text{ hours a candle burns}$$
$$\text{DV (Output):} \quad H(t) = \text{ height of the candle in cm}$$

Using H as a name for the function, we can write the ordered pairs (1 hr, 28 cm) and (10 hrs, 22 cm) as $H(1 \text{ hr}) = 28$ cm and $H(10 \text{ hrs}) = 22$ cm.

For the function **(c)** the independent and dependent variables are

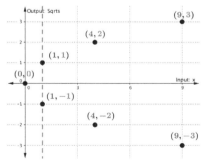

Figure 1.18 Square Roots **(b)**.

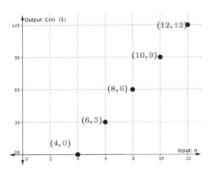

Figure 1.19 Toppings **(c)**.

IV (Input): n = **number of toppings**
DV (Output): $C(n)$ = **total cost of toppings in $**

Using C as a name for the function, we can write the ordered pairs (4, $0)
and (8, $6) as $C(4)$ = $0 and $C(8)$ = $6.

Given a graph of a relation, to decide whether the relation is a function or not,
we need to check whether each input has exactly one output. Notice when you
graph two points that have the same input and different outputs, they lie one
above the other on the same vertical line as all the points on a vertical line have
the same input. Thus if we can find a vertical line containing at least two points
of the graph, then we get one input having more than one output and the relation
is not a function. If such a line cannot be found, the relation is a function.

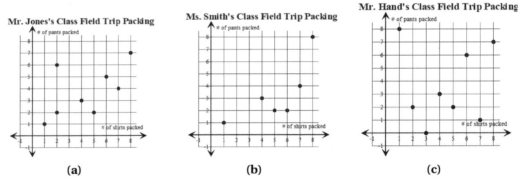

Figure 1.20 Is it a Function?

IV (In)	DV (Out)
x (shirts)	J (pants)
1	1
2	2
2	6
4	3
5	2
6	5
7	4
8	7

Example 1.12 (a)

Example 1.12 For each graph above make a table to represent the relation.
Determine whether or not it represents a function and EXPLAIN why.

IF it IS a function:

Identify and label the independent and dependent variables and give the
function a name.

Then express at least two of the ordered pairs in function notation.

Solution: The tables are displayed in the margin. The relation **(a) is not
a function** since the points (2, 2) and (2, 6) on the graph have the same
input: 2, but two outputs: 2 and 6. The relations **(b)** and **(c) are functions**
since each input has exactly one output.

For the function **(b)** the independent and dependent variables are

IV (Input): x = **number of shirts**
DV (Output): $S(x)$ = **number of pants for Smith's class**

Using S as a name for the function, we can write the ordered pairs $(1, 1)$ and $(4, 3)$ as $S(1) = 1$ and $S(4) = 3$.

For the function **(c)** the independent and dependent variables are

IV (Input): $x =$ **number of shirts**
DV (Output): $H(x) =$ **number of pants for Hand's class**

Using H as a name for the function, we can write the ordered pairs $(2, 2)$ and $(5, 2)$ as $H(2) = 2$ and $H(5) = 2$.

IV (In) x (shirts)	DV (Out) $S(x)$ (pants)
1	1
4	3
5	2
6	2
7	4
8	8

Example 1.12 (b)

Wrapping Up

We have seen that paring two related variables leads to relations as collections of ordered pairs of the form **(input, output)**. The set of all inputs is the domain of the relation and the set of all outputs is the range of the relation. Functions are special relations with the property that each input in the domain matches *exactly one* output in the range.

As collections of ordered pairs, functions can be represented by **tables** listing these pairs as rows **(input, output)** or by **graphs**, having these pairs plotted as points in a plane with coordinates **(input, output)**. The input is also called the independent variable as it changes independently, and the output is called the dependent variable as its values depend on the values of the independent variable.

As a *rule* which assigns to each input *exactly one* output, a function can be given a name, say f, and the output can be relabeled by using the function notation **output** $= f($**input**$)$. This notation allows us to identify *both* coordinates of a point **(input, output)**. In the next section, we introduce yet another representation of a function as a set of ordered pair solutions to an equation in two variables. These equations are used to model functions in the real world.

IV (In) x (shirts)	DV (Out) $H(x)$ (pants)
1	8
2	2
3	0
4	3
5	2
6	6
7	1
8	7

Example 1.12 (c)

Exercises

Exercises for 1.4 Relationships between Variables - Introduction to Functions

P1.64   For each set of ordered pairs below, determine the domain and range. Then specify whether it is a function or not and justify your answer. IF it IS a function, give it a name and write at least two ordered pairs in function notation.

      a) {(9,11), (11,4), (2,−5), (−6,8), (2,6), (17,0)}

      b) {(Ford, blue), (Nissan, red), (Toyota, silver), (Subaru, black), (Chevy, blue), (Volvo, red)}

P1.65   For each relation shown in **Fig 1.21**:

      a) Identify the domain and range.

      b) Determine whether or not it is a function and explain your reasoning.

      c) If it is a function, give it a name and write two of its pairs using function notation.

Suit	Color
Hearts	Red
Spades	Black
Diamonds	Red
Clubs	Black

(I)

d (day)	S (special)
Monday	Burgers
Tuesday	Chili
Wednesday	Chicken
Thursday	Burgers
Friday	Fried Fish

(II)

Flower	Color
Rose	Red
Carnation	Yellow
Tulip	Red
Orchid	Purple

(III)

Input	Output
Δ	♀
Σ	∞
☺	☼
π	◊

(IV)

Figure 1.21

Exercises for 1.4 Relationships between Variables - Introduction to Functions

P1.66   For each relation shown in **Fig 1.22**:

      a) Identify the domain and range.

      b) Graph the relation on a coordinate grid.

      Label axes and scales.

      c) Determine whether or not it is a function and explain your reasoning.

      d) If it is a function, give it a name and write two of its pairs using function notation.

Temperature (°C)	Volume of Gas (mL)
20	60
40	65
60	70
80	75
100	80
120	85

(I)

age	# visits to Disneyland
2	0
5	1
9	3
12	2
9	4
11	2

(II)

# folds	# layers thick
0	1
1	2
2	4
3	8
4	16
5	32

(III)

t = Time Traveled (hrs.)	D(t) = Distance Driven (km.)
2	220
3	320
4	420
7.5	770

(IV)

Figure 1.22

Exercises for 1.4 Relationships between Variables - Introduction to Functions

P1.67   For each relation shown in **Fig 1.23**:

    a) Make a table to represent the relation.

    b) Give the domain and range of the relation.

c) Determine whether or not the relation is a function and explain your reasoning.

d) If it is a function, express two of its points using function notation.

e) If it is a function, give an example of a point that if added to the plot would make it NOT a function. Explain.

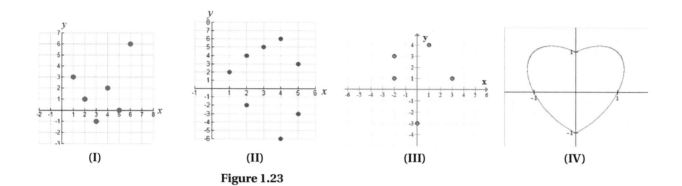

(I)   (II)   (III)   (IV)

Figure 1.23

1.5   Analyzing Real World Functions (Models)

Launch Exploration

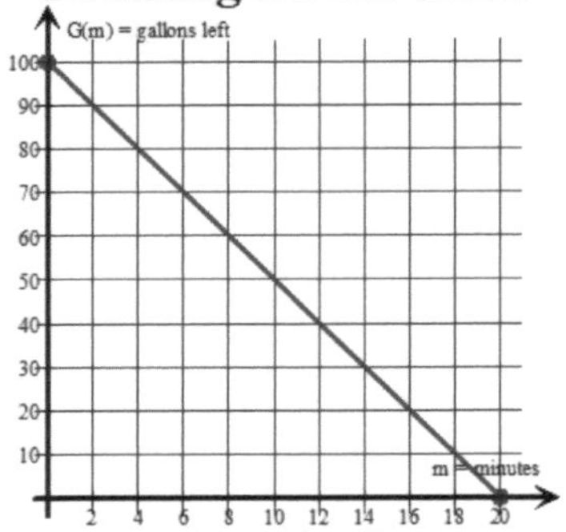

Draining a Fish Tank

Figure 1.24 Find ordered pairs.

Sometimes a relationship between the input and output of a function can be described by an equation. For example, when draining a fish tank the amount of water left depends on the draining time. In this situation, we choose

IV (Input): $m =$ **amount of draining time (minutes)**
DV (Output): $G(m) =$ **amount of water left in the tank (gallons)**

The two variables are related by the equation $G(m) = 100 - 5m$ which allows us to find the output $G(m)$ given the input m. For example, if the input is $m = 10$ minutes, then the output is $G(10) = 100 - 5(10) = 50$ gallons and the ordered pair $(10, 50)$ can be listed in a **table** and plotted in a **graph**. Using both the equation AND the graph **Fig 1.24**, answer the questions below. Express each answer as an ordered pair and in function notation. Then enter the ordered pair in a table.

- How much water is in the tank after **6** minutes of draining?

- After how many minutes of draining will there be **40** gallons left in the tank?

- Find $G(5)$ and explain its meaning in the context of the situation.

- Find m when $G(m) = 35$ and interpret its meaning.

Function Equations

Functions in the real world can often be represented by equations in two variables. In our **Launch Exploration** the equation is

$$(1.5) \qquad\qquad G(m) = 100 - 5m$$

where m is the independent variable and $G(m)$ is the dependent variable. To find ordered pairs of the form **(input, output)**, we choose input values for m and find their output values $G(m)$ by substitution and evaluation. For example, we substitute the input $m = 6$ **min** into the **Equation (1.5)** and evaluate the output:

(Substitute & Evaluate) $(1.6) \qquad\qquad G(6 \text{ min}) = 100 - 5(6) = 70 \text{ gal}.$

Thus the ordered pair $(6, 70)$ is a *solution* to the *function equation* (**1.5**) and it means that after **6 min** of draining there will be **70 gal** left in the tank.

Fact 1.14 *Function Equation (Formula)*

A function can be given as a set of **ordered pair solutions** to an **equation (formula)** in two variables of the form $f(\text{input}) = \text{output}$.

Finding Ordered Pair Solutions

IV (In) m (min)	DV (Out) $G(m)$ (gal)
0	100
1	95
5	75
6	70
12	40
13	35
17	15
20	0

Table 1.16 Ordered pair solutions.

*After how many minutes of draining will there be **40** gallons left in the tank?*

We are given the output $G(m) = 40$ gal and we have to find the input m that gives that output by solving the equation below for m:

$$G(m) = 100 - 5m$$
$$40 \text{ gal} = 100 - 5m$$
$$-60 = -5m$$
$$12 \text{ min} = m.$$

We enter the ordered pair solution (**12**, **40**) in **Table 1.16**.

Answer. After **12 min** of draining there will be **40 gal** left in the tank.

*Find **G(5)** and explain its meaning in the context of the situation.*

We are given the input $m = 5$ **min** and to calculate the output we substitute the input in the function equation and evaluate the output:

$$G(5 \text{ min}) = 100 - 5(5) = 100 - 25 = 75 \text{ gal}.$$

We enter the ordered pair (**5**, **75**) in **Table 1.16**.

Answer. After **5 min** there are **75 gal** left in the tank.

*Find **m** when **G(m)** = **35** and interpret its meaning.*

As before, we are given the output $G(m) = 35$ **gal** and we find the input by solving the function equation for m:

$$G(m) = 100 - 5m \qquad \text{(Function Equation)}$$
$$35 \text{ gal} = 100 - 5m \qquad \text{(Substitute)}$$
$$-65 = -5m \qquad \text{(Subtract 100)}$$
$$13 \text{ min} = m. \qquad \text{(Divide by } -5)$$

We enter the ordered pair solution (**13**, **35**) in **Table 1.16**.

Answer. After **13 min** there are **35 gal** left in the tank.

Function Graphs

The ordered pair solutions to the function **Equation (1.5)** can be listed in the function **Table 1.16** and plotted on the function graph **Fig 1.24**. Conversely, we can use the graph to estimate the answers to the previous questions with various degrees of accuracy. For example, the points (**6**, **70**) and (**12**, **40**) are accurately estimated as intersections of gridlines (**lattice points**) while the points (**5**, **75**) and (**13**, **35**) are less accurately estimated. The only way to check their accuracy is to substitute their coordinates into the function equation to make sure it is a true statement. Moreover, using the graph we can answer additional questions.

Reading Ordered Pairs from a Graph

After how many minutes will the tank be completely drained?

The tank is empty when $G(m) = 0$ gallons left. We must find m when $G(m) = 0$. We look at the graph for a point with second coordinate 0: (20, 0). In function notation we write $G(20) = 0$.

Answer. It takes 20 minutes to drain the tank completely.

This is called the '*x*'-*intercept* (*m*-*intercept*) for this function which is an ordered pair with output 0. Using **Equation 1.5**, we check $G(20) = 100 - 5(20) = 0$.

How much water was in the tank when the draining started?

Before the draining starts $m = 0$ minutes of draining. We can look at the graph for a point on it with a first coordinate of 0: (0, 100). So $G(0) = 100$.

Answer. There were initially (at the start, before draining) 100 gallons in the tank.

This is the '*y*'-*intercept* ($G(m)$-*intercept*) for this function which is an ordered pair with input 0. Using **Equation 1.5**, we can check $G(0) = 100 - 5(0) = 100$.

Identify at least two additional data points from the graph. Write them using function notation and as ordered pairs. Interpret their meaning in context.

Choose any (valid) value for the input m, say $m = 1$ min or $m = 17$ min, and look at the graph for a point with the first coordinate being that input:

Answer. The data point (1, 95) or $G(1) = 95$ means that there are 95 gal left after 1 min. The point (17, 15) or $G(17) = 15$ means that there are 15 gal left after 17 min.

We can verify that all data points in **Table 1.16** satisfy the function **Equation (1.5)** and we can plot them in **Fig 1.25**.

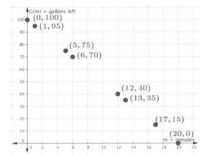

Figure 1.25 Scatter Plot.

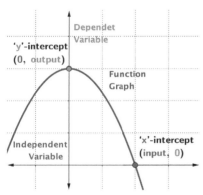

Figure 1.26 Intercepts.

Definition 1.15 *The Intercepts of a Function*

An '*x*'-**intercept** of a function f is a point **(input, 0)** where the graph crosses the horizontal axis. Its '*x*'-**coordinate** is an input x with the output $f(x) = 0$. The '*y*'-**intercept** is the point **(0, output)** where the graph crosses the vertical axis. Its '*y*'-**coordinate** is the output $y = f(0)$ for the input $x = 0$.

The intercepts of a function contain important information about the function. For example, in the fish-tank problem the '*x*'-*intercept* is named *m*-*intercept* and its *m*-*coordinate* is the time needed to empty the tank. The '*y*'-*intercept* is named $G(m)$-*intercept* and its $G(m)$-*coordinate* is the initial amount of water in the tank. A function cannot have more than one '*y*'-intercept since one input (0) has only one output, but it can have multiple '*x*'-intercepts. See **Fig 1.26**.

Functions as Real-World Models

Functions and their various representations are used as mathematical **models** for many aspects of reality involving a relation between two variables. These models are **valid** only for certain values of the variables.   For example, if we choose $m = 22$ minutes in the **Launch Exploration**, then

(**1.7**) $G(22) = 100 - 5(22) = -10$ gallons.

But that value does not make sense in this situation as the tank cannot have a negative amount of water in it. In other words, -10 is not in the range. Since the value $m = 22$ minutes gives us an invalid output, $m = 22$ is an invalid input and is not in the domain. In fact, the tank was empty after 20 minutes. So the function is *not a valid model for $m < 0$ or $m > 20$* because we do not start to drain the tank before $m = 0$ and we cannot continue to drain it after $m = 20$. Since the input variable m is the draining time and thus, it is a *continuous* variable, the *real-world domain* of the **fish-tank function G** is the interval of ALL real numbers from 0 to 20 (inclusive) and we write

Real World Domain of G = [0 min, 20 min].

In this time interval, the amount of water in the tank *decreases* from 100 gallons when the tank is full to 0 gallons when the tank is empty. Since the output variable $G(m)$ is the amount of water and thus a *continuous* variable, that means that the *real-world range* of the function G is the interval of ALL real numbers from 0 to 100 (inclusive) and we write

Real World Range of G = [0 gal, 100 gal].

To draw the graph of the function G we start with a scatter plot as in **Fig 1.25**, making sure to choose scales for our axes that allow us to include the endpoints of our domain and range. The scatter plot shows the **trend** of the function is steadily *decreasing* and since the variables are *continuous*, **all** the points on this trend must belong to the graph. Therefore the complete **graph** of the function G is the **solid** line which connects the points of the scatter plot. See **Fig 1.27**.

Definition 1.16 *Increasing or Decreasing Functions*
A function is **decreasing** or **increasing** if its output values get respectively smaller and smaller or bigger and bigger as the input values get bigger and bigger. The graph of a decreasing or increasing function goes respectively DOWN or UP from LEFT to RIGHT.

Notice that the graph in **Fig 1.27** is decreasing and can be extended beyond the real-world domain by ordered pair solutions to the **Equation** (**1.5**) such as $(-2, 110)$ or $(22, -10)$ shown in red, but these points do not belong to the graph of G since they do not have real-world interpretation. We can say that these points belong to the *mathematical* graph of the function equation.

Interval	Notation
All real numbers	$(-\infty, +\infty)$
$x < 2$	$(-\infty, 2)$
$x \leq 2$	$(-\infty, 2]$
$-5 < x < 3$	$(-5, 3)$
$-5 \leq x \leq 3$	$[-5, 3]$
$1 \leq x$	$[1, +\infty)$
$1 < x$	$(1, +\infty)$

Table 1.17

☞ In the interval notation $[L, R]$ the left end L must be less than the right end R.

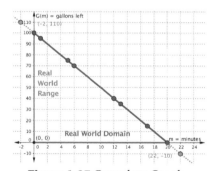

Figure 1.27 Complete Graph.

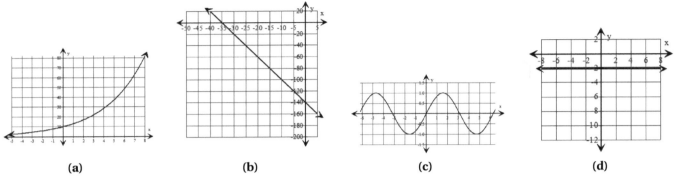

(a)   **(b)**   **(c)**   **(d)**

Figure 1.28 Is it Increasing or Decresing?

In the graphs above we have **(a)** an increasing function, **(b)** a decreasing function, **(c)** an oscillating function which is neither increasing nor decreasing *throughout* the domain, and **(d)** a constant function which is neither increasing nor decreasing in *any part* of the domain. A **constant function** is a function which assigns the same output value to every input value. For example, $f(x) = 4$ assigns the same output value 4 to each input value x.

The Initial Value of a Function

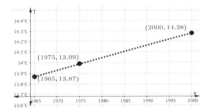

Figure 1.29 Initial Value in 1965.

The **value of a function** f at an input value x is simply the output value $f(x)$ assigned to that input. For example, the value of the constant function $f(x) = 4$ is the constant 4, a quantity that does not change with the input. The value of the function given by the formula $G(m) = 100 - 5m$ in the **Launch Exploration** at $m = 10$ is 50 as $G(10) = 100 - 5(10) = 50$ by *evaluation*. The value of the same function at $m = 0$ is 100 as $G(0) = 100$ and this value represents the *initial* amount of water in the fish-tank before starting the draining. In general, *we define the* **initial value** *of a function* f *to be the output value of* f *at the input of zero,* $f(0)$.

In applications, we may have data starting at a nonzero input. For example, the average global temperature in 1965 was $13.87°$C and the average temperature in the year y *after* 1965 is predicted by the formula $T(y) = 0.0116y - 8.9214$. The initial value for this formula is given in the year $y = 1965$ by

$$T(1965) = 0.0116(1965) - 8.9214 \approx 13.87°\text{C}.$$

The average temperature in 1975 is predicted to be $T(1975) \approx 13.99°$C. However, if we consider the input variable to be the *number of years, t, since* 1965, the new formula for the average temperature t years after 1965 is

$$C(t) = 0.0116t + 13.87.$$

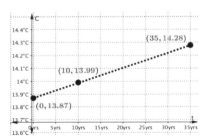

Figure 1.30 Initial Value at 0.

This formula gives the *exact* initial value $C(0) = 13.87°$C as 1965 corresponds to $t = 0$. Similarly, 1975 is $t = 10$ years after 1965 and its average temperature is

$$C(10) = 0.0116(10) + 13.87 \approx 13.99°\text{C}.$$

The graphs of the two formulas are given in the margin **Figures 1.29** and **1.30**.

Four Ways to Represent Functions

For a complete picture of a function the following representations should be worked out together (**the rule of four**):

- In words - verbal description of the rule between input/outputs (**verbal**)

- In a table/list of ordered pairs (**table**)

- In a graph or scatter plot with axes having appropriate scales (**graph**)

- As an equation - a symbolic description (**equation**)

A good strategy for working with functions is to start a table of ordered pairs that you can fill in as you answer questions. Do this even if the 'instructions' do not tell you to make a table. See **Table 1.18** in the margin.

> **Example 1.13** To raise money, a campus club buys 600 t-shirts to sell. The profit they earn is a function of the number of shirts they sell and is given by the function equation: $P(s) = -2400 + 10s$, where s is the number of shirts sold, and $P(s)$ is the profit earned in dollars.
>
> Do the analysis suggested in **Table 1.18** and represent this function in words, in a table, and in a graph. Answer the following questions using both ordered pairs and function notation:
>
> **a)** How much profit does the club earn if they sell 400 shirts?
>
> **b)** Can the club have a profit of −$1650? Explain.
>
> **c)** What is the maximum profit they can earn? Explain.

Solution: The profit depends on the number of shirts sold.

> IV (Input): $s = $ **the number of shirts sold**
> DV (Output): $P(s) = $ **the profit earned from selling shirts ($)**

They cannot sell a partial shirt so s is a **discrete** variable - it can only take non-negative integer values. The profit in dollars is also **discrete** taking on values only in separate dollar amounts.

Verbal Representation

The profit in dollars is $2400 less than ten times the number of shirts sold.

Table Representation

Keys to Function Analysis

Identify and label the independent and the dependent variables.

Determine whether the variables are continuous or discrete.

Start a table to list some ordered pairs including the intercepts.

Determine and write the real-world domain and range.

Draw a graph or scatter plot with appropriate scales.

Determine whether the function is increasing or decreasing.

Table 1.18

In **Table 1.19**, the 'x'-**intercept** is the point with an output value of 0. For this function it is the s- intercept with $P(s) = 0$. We solve for s:

(Substitute)

$$P(s) = -2400 + 10s, \qquad\qquad 0 = -2400 + 10s,$$

(Solve for s)

$$2400 = 10s, \qquad\qquad 240 = s.$$

So the s-**intercept** is $P(240) = 0$ or the point $(240, 0)$. This means when the club sells **240 shirts** they earn a profit of **$0**. This is the break-even point when they do not lose or gain money.

The 'y'-**intercept** is the point with an input value of 0. For this function it is the $P(s)$-intercept with $s = 0$.

$$P(0) = -2400 + 10(0) = -2400 + 0 = -\$2400.$$

IV (In)	DV (Out)
s (shirts)	$P(s)$ ($ profit)
0	−2400
75	−1650
100	−1400
240	0
300	600
400	1600
600	3600

Table 1.19 Table Representation.

The $P(s)$-**intercept** is $P(0) = -2400$ or $(0, -2400)$. This means if the club does not sell **any shirts** they will have a loss of **$2400**.

a) *How much profit does the club earn if they sell 400 shirts?*

We are given $s = 400$ shirts so we evaluate the function at $s = 400$:

$$P(400) = -2400 + 10(400) = -2400 + 4000 = \$1600.$$

In function notation we write $P(400) = \$1600$ and have the point $(400, 1600)$. They earn **$1600** in profit if they sell 400 shirts.

b) *Can the club have a profit of −$1650? Explain.*

We check by substituting $P(s) = -1650$ into the equation and solve for s:

(Substitute)

$$P(s) = -2400 + 10s, \qquad\qquad -1650 = -2400 + 10s,$$

(Solve for s)

$$750 = 10s, \qquad\qquad 75 = s.$$

Yes, the club bought 600 shirts to sell so they could sell only 75 that would make −$1650 in profit which means they lose money. In function notation we have $P(75) = -1650$. The point is $(75, -1650)$.

c) *What is the maximum profit they can earn? Explain.*

The maximum profit will come if they sell all 600 shirts. So we evaluate the function at $s = 600$:

$$P(600) = -2400 + 10(600) = -2400 + 6000 = \$3600.$$

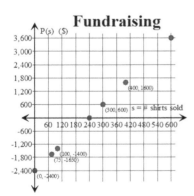

Figure 1.31 Scatter Plot.

So $P(600) = \$3600$ is the maximum possible profit. The point is $(600, 3600)$.

In **Table 1.19**, we also list the answers above and a few extra data points obtained from the function equation.

Graph Representation

Since the number of shirts sold is from 0 to 600 and the profit earned is from −2400 to 3600, we have

Real-World Domain = { 0 shirts, 1 shirt, 2 shirts, ... , 600 shirts}

Real-World Range = {−$2400, −$2390, −$2380, ..., $3600}

To show the endpoints of the domain and range in a **scatterplot**, we scale the input axis by ticks 60 shirts apart and the output axis by ticks $600 apart. Then we plot the sample of points from **Table 1.19** in **Fig 1.31**. This shows an **increasing** trend for the function.

In **Example 1.13** we consider the variables to be *discrete* and in this case the complete **graph** of the function P is the **dotted** line connecting the data points in the scatterplot. See **Fig 1.32**. The scatterplot consists of a sample of 7 points and the complete graph consists of all 601 points corresponding to the 601 input values in the domain. Although the line segment can be extended beyond its endpoints, the model cannot be extended to a larger domain.

Wrapping Up

We have seen that a function can be used as a **model** of a real-world situation and can be represented as a set of ordered pair solutions to an **equation** in two variables. This equation allows us to start an input-output **table** and to make a **scatter plot** for a sample of data points of the function. This scatter plot may have an **increasing** or **decreasing** trend and we can connect its points by a solid line if the variables are *continuous* or by a dotted line if the variables are *discrete*. Not every input gives a valid output for the model and the **real-world domain** and **range** are often intervals or finite sets of real numbers for which the model makes sense. In particular, the points where the **graph** of the function intersects the axes are called **intercepts** and often contain important information. For that reason, sometimes the term 'y-intercept' is used for the **initial value** of the function, $y = f(0)$. In the remainder of the book we will study the linear, exponential, and quadratic patterns in data that allow us to write function equations (formulas) and make predictions. These models cover a lot of applications in physics, finances, trade, commerce, biology, etc.

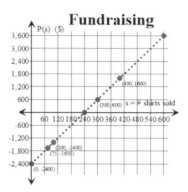

Figure 1.32 Complete Graph.

☞ For a very large number of data points, we can treat the variables as continuous and use a solid line graph.

Exercises

Exercises for 1.5 Analyzing Real World Functions (Models)

P1.68   Analyze the graph and answer the question

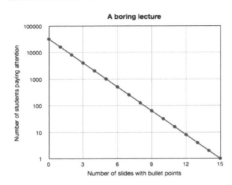

P1.69   The graph shows the length $L(w) = 2 + w$ of a certain spring when different weights w are attached to it.

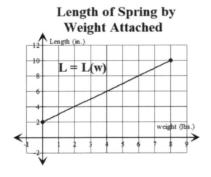

Find the missing value in the ordered pair described by the function notation and explain the meaning of that ordered pair in context including units.

   a) $L(2)$

   b) $L(w) = 2$

   c) $L(w) = 4$

   d) $L(4)$

   e) $L(10)$

   f) $L(w) = 8$

P1.70   The graph of the function $A(w) = 950 - 50w$ is shown below.

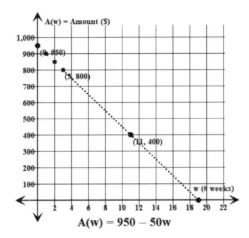

   a) Create a story around this function. What do you think is going on?

   b) Explain the meaning of $A(18)$ including units and use the equation to find the other value in the ordered pair.

   c) Explain the meaning of $A(w) = 300$ including units and use the equation to find the other value in the ordered pair.

   d) Use the function equation to find the amount of money after 8 weeks. Express the answer as an ordered pair and find the point on the graph.

   e) Use the function equation to find the number of weeks when the amount is $450. Express the answer as an ordered pair and find the point on the graph.

   f) Give an example of a question about the money or the weeks that does not make sense for this function. Explain why it doesn't make sense.

In problems **P1.71** to **P1.75**, the graph and the function equation are given. Start a table for each and enter the ordered pairs labeled on the graph. Then add any new ordered pairs to the table as you answer the questions.

P1.71   Suppose you want to plant a rectangular garden with an area of 100 square feet. If we label the south side length s and the west side length w, then the area of the garden is $100 = sw$, and solving for w we get the function $w = 100/s$.

Side Lengths of a Rectangle
with Area = 100 square feet

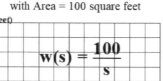

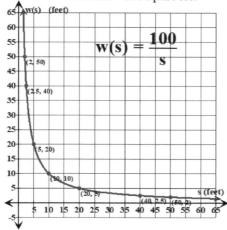

a) What kind of garden do you think this is? Flowers? Herbs? Vegetables?

b) Use the function equation to find the south side length of the garden if the west side is 12 feet. Round your answer to the tenths. Express the answer as an ordered pair and find the point on the graph.

c) Use the function equation to find the west side of the garden if the south side is 30 feet. Express the answer as an ordered pair and find the point on the graph.

d) How does the shape of the garden in question **b**) compare to the garden in **c**)?

e) Can either side of the garden have a length of 0 feet? Explain.

f) What dimensions do you think would be best for the garden? Explain why.

P1.72   Minerva is choosing tile for her home improvement project. The function $S(a) = \sqrt{a}$ gives the side length $S(a)$ of a square tile with area a.

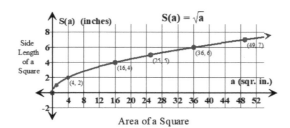

a) What do you think Minerva is tiling? What size square tile do you think she should use? Why?

b) Explain the meaning of $S(4)$ including units and use the equation to find the other value in the ordered pair.

c) Explain the meaning of $S(a) = 4$ including units and use the equation to find the other value in the ordered pair.

d) Use the function equation to find the side length of a square tile with an area of 32 square inches. Round your answer to the tenths. Express the answer as an ordered pair and find the point on the graph.

e) Use the function equation to find the area of a square tile with a side length of 6.5 inches. Express the answer as an ordered pair and find the point on the graph.

P1.73   The function $D(h) = 60h$ gives the total distance, $D(h)$ in miles, traveled by friends on a road trip after h hours of driving.

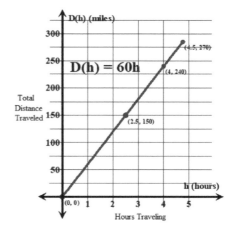

a) Where do you think the friends are going? Where would you want to go with your friends on a road trip?

b) Explain the meaning of $D(3/4)$ including units and use the equation to find the other value in the ordered pair.

c) Use the function equation to find the distance the friends traveled in 150 minutes. Express the answer as an ordered pair and find the point on the graph.

d) Use the function equation to find the amount of time the friends have been traveling when they have traveled 80 miles. Express the answer as an ordered pair and find the point on the graph. How much time is that in minutes?

P1.74 The function $C(n) = 10 + 1.5n$ gives the total cost of the amusement park, $C(n)$, in dollars depending upon the number of rides taken, n.

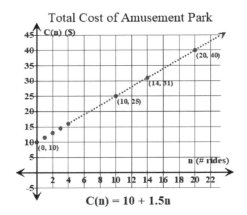

Total Cost of Amusement Park

$C(n) = 10 + 1.5n$

a) Explain the meaning of $C(20)$ including units and use the equation to find the other value in the ordered pair.

b) Explain the meaning of $C(n) = 16$ including units and use the equation to find the other value in the ordered pair.

P1.75 The function $h(t) = -16t^2 + 50t + 5$ gives the height, $h(t)$, in feet, of a thrown ball t seconds after it is released.

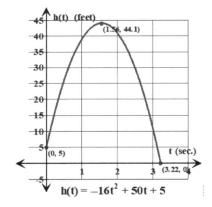

$h(t) = -16t^2 + 50t + 5$

a) Verify that the point $(1, 39)$ is a point on the graph and a solution to the function equation. Express the point $(1, 39)$ in function notation.

b) Explain the meaning of the point $(1, 39)$ in the context of the situation including units.

c) Explain the meaning of the point on the graph labeled $(1.56, 44.1)$. What is special about this point?

d) Use the function equation to find the coordinates of another point on the graph. Express it in function notation and explain its meaning in context.

P1.76 Review each problem from **P1.71** to **P1.75** and complete the following tasks:

a) Are the variables discrete or continuous? Explain how you know.

b) Is the function increasing, decreasing, or neither?

c) What is the initial value of the function? Explain what it means in context and express that point as an ordered pair and in function notation.

d) Find the $'x'$-intercept and explain its meaning in context. Include units.

e) Give the domain and range of the function including units. Use correct notation depending on whether variables are continuous or discrete.

f) Identify a point labeled on the graph and verify that it is a solution to the function equation. Express it using function notation. Explain what it means in context.

g) Use the equation to find a point on the graph that is not already labeled. Explain what it means in context.

P1.77 For each function, (1) sketch the graph by plotting points, (2) find its intercepts if any, and (3) determine the domain for which the formula makes sense:

a) $f(x) = 1/x$

b) $g(x) = |x|$

c) $h(x) = \sqrt{x}$

Unit 2

Linear Models

A Review of Main Ideas

Before we begin this section, let us summarize a few facts:

- A relation is any pairing of numbers. A function is a special type of relation that matches each allowable input to one and only one output.

- The independent variable (IV) is the input variable. It is the variable that is independently changing.

- The dependent variable (DV) is the output variable. It is the variable that is affected by the change in the independent variable.

- The real-world domain for a function (model) is the set of all inputs that make sense for that model. The range is the set of all outputs.

- A function (model) can be described verbally or represented by tables, graphs, or equations.

2.1   Introduction to Linear Relationships

Launch Exploration

Linda bought a seedling plant that is one foot tall and it will grow about 4 inches every 3 weeks for the first 15 weeks.

What should we ask first? What are the variables in this situation? How many variables are in a function? Which one is the independent variable and which one is the dependent variable? We must choose appropriate units for each variable. How tall will the plant be after 15 weeks? Get together with a group of classmates and do the tasks in **Table 2.1** on the margin except for the last one. In other words, study this relationship with no equation in sight.

Keys to Modeling

Identify the dependent vs. independent variables including units.

Make a table of values.

Examine the table for a pattern.

Identify the real-world domain and range.

Create a graph for the function.

Write an equation (formula) to represent the function.

Table 2.1

Linear Function Tables

Let us choose the time since the plant was bought as the **independent variable** and the height of the plant as the **dependent variable**. We label the variables including the units below:

IV (Input): t = the time since the plant was bought (**unit:** weeks)
DV (Output): $h(t)$ = the height of the plant (**unit:** inches)

Next we draw the plant every 3 weeks and make a **table of values** for the height by successively adding 4 inches to the initial height of 1 foot. To do the calculations using a common unit, we convert 1 foot into 12 inches.

Figure 2.1 Linda's Plant

Diff	t	$h(t)$	Diff
+3	0	12	+4
+3	3	16	+4
+3	6	20	+4
+3	9	24	+4
+3	12	28	+4
	15	32	

Table 2.2 Linda's Plant Table

We record these data points in **Table 2.2** on the margin. What is the pattern? Observe that *the differences between outputs are constant as the inputs increase by a constant.* This pattern characterizes the *Linear Function Tables.*

Definition 2.1 *Linear Function Tables*

Linear function tables are tables of values characterized by **constant differences** of the outputs *when the inputs increase by a constant.*

To draw the graph of the relation between $h(t)$ and t we need to determine the real-world domain. Since this model is valid for 15 weeks and both variables are *continuous* (take on in-between values), we have

(2.1)   **Domain = [0 weeks, 15 weeks],   Range = [12 inches, 32 inches],**

where the lowest output value $h(0) = 12$ inches and the highest output value $h(15) = 32$ inches are read off from the table. In **Fig 2.2** the distance between ticks represents 3 weeks on the t-axis and 4 inches on the $h(t)$-axis. Notice that the plotted points line up in the graph and so, **the graph** is the solid straight line joining the points $(0, 12)$ and $(15, 32)$. This justifies the terminology *linear* graph or function. The line can be extended beyond the endpoints to show the trend, but that extension is not part of the real-world graph.

The constant difference takes us from output to output going forward in the linear **Table 2.2**

(2.2)   **current output + difference = next output**

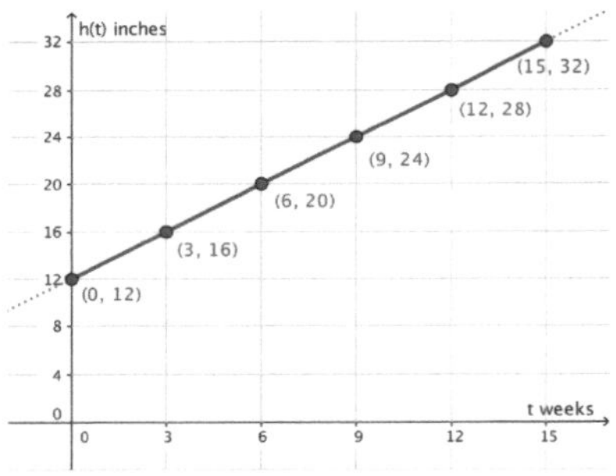

Figure 2.2 Linda's Plant Graph.

and it is obtained by taking the differences of outputs going backwards

(2.3)   difference = **current output − previous output**.

☞ Δ (Delta) is the Greek letter for D as in "Difference".

First Differences Notation:

$(\Delta x)_1 = x_1 - x_0$

$(\Delta x)_2 = x_2 - x_1$

...

$(\Delta y)_1 = y_1 - y_0$

$(\Delta y)_2 = y_2 - y_1$

...

Definition 2.2 *First Order Differences*

In a table of values of a function $y = f(x)$, the **first (order) differences** Δy and Δx respectively of the **output y** and the **input x** are defined by

1st Diff (Δx)	x	y	1st Diff (Δy)
$x_1 - x_0$	x_0	y_0	$y_1 - y_0$
$x_2 - x_1$	x_1	y_1	$y_2 - y_1$
$x_3 - x_2$	x_2	y_2	$y_3 - y_2$
	x_3	y_3	

These differences represent Δx = **change in x** and Δy = **change in y**.

Another Plant Example

Cheryl bought a seedling plant that is one foot tall and it will grow about 5 inches every 4 weeks for the first 16 weeks. Let us label the variables including the units:

IV (Input): t = the time since the plant was bought (**unit:** weeks)

DV (Output): $g(t)$ = the height of the plant (**unit:** inches)

Next we draw the plant every 4 weeks and make a **table of values** for the height by successively adding 5 inches to the initial height of 1 foot. To do the calculations, we convert 1 foot into 12 inches.

Diff	t	$g(t)$	1st Diff
+4	0	12	+5
+4	4	17	+5
+4	8	22	+5
+4	12	27	+5
	16	32	

Table 2.3 Cheryl's Plant Table

0 weeks   4 weeks   8 weeks   12 weeks

Figure 2.3 Cheryl's Plant.

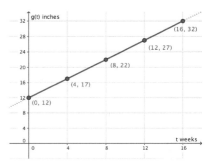

Figure 2.4 Cheryl's Plant Graph.

We record these data points in **Table 2.3** on the margin. Again, observe that *the first differences between outputs are constant as the inputs increase by a constant.* Since the model is valid for 16 weeks and both variables are *continuous*, we have

(2.4)   **Domain = [0 weeks, 16 weeks],   Range = [12 inches, 32 inches],**

> ☞ A point (x_1, y_1) of a function (graph) is a solution to the function equation $y = f(x)$ or a row in the table.

where the lowest output value $g(0) = 12$ inches and the highest output value $g(16) = 32$ inches are read from the table. In **Fig 2.4** the distance between ticks represents 4 weeks on the t-axis and 4 inches on the $g(t)$-axis. Notice that the plotted points line up again on the graph and so, the graph is the solid straight line joining the points $(0, 12)$ and $(16, 32)$.

Constant Rates of Change

Given the two situations above, which plant is growing faster? Linda's or Cheryl's? To answer this question we introduce

> ☞ The rate of change of y with respect to x is $\Delta y / \Delta x$, NOT $\Delta x / \Delta y$.
>
> If $\Delta y = y_2 - y_1$ then we use $\Delta x = x_2 - x_1$, NOT $x_1 - x_2$.

Definition 2.3 *The Rate of Change*

Given two ordered pairs or **points** (x_1, y_1) and (x_2, y_2) of a function $y = f(x)$, the **rate of change in y with respect to x as the input x changes from x_1 to x_2** is the ratio of the change in y to the change in x

$$\textbf{rate of change} = \frac{\textbf{change in } y}{\textbf{change in } x} = \frac{\textbf{Diff } y}{\textbf{Diff } x} = \frac{\Delta y}{\Delta x} = \frac{y_2 - y_1}{x_2 - x_1}.$$

For Linda's plant we can use the points $(x_1, y_1) = (0, 12)$ and $(x_2, y_2) = (3, 16)$ to get the rate of change

Diff $= \Delta t$	t	$h(t)$	Diff $= \Delta h$
$+3$	0	12	$+4$
	3	16	

Linda's Plant Table

(2.5)   Linda's rate $= \dfrac{\Delta h}{\Delta t} = \dfrac{16 \text{ in} - 12 \text{ in}}{3 \text{ wk} - 0 \text{ wk}} = \dfrac{4 \text{ in}}{3 \text{ wk}} = \dfrac{4/3 \text{ in}}{1 \text{ wk}} \approx 1.33$ in. per wk.

For Cheryl's plant we can use the points $(x_1, y_1) = (0, 12)$ and $(x_2, y_2) = (4, 17)$ to get the rate of change

(2.6)   Cheryl's rate $= \dfrac{\Delta g}{\Delta t} = \dfrac{17 \text{ in} - 12 \text{ in}}{4 \text{ wk} - 0 \text{ wk}} = \dfrac{5 \text{ in}}{4 \text{ wk}} = \dfrac{5/4 \text{ in}}{1 \text{ wk}} = 1.25$ in. per wk.

> ☞ Linda's plant rate using $(0, 12)$ and $(15, 32)$:
>
> $\frac{\Delta h}{\Delta t} = \frac{32 - 12}{15 - 0} = \frac{20}{15} = \frac{4}{3}$.
>
> Cheryl's plant rate using $(0, 12)$ and $(16, 32)$:
>
> $\frac{\Delta g}{\Delta t} = \frac{32 - 12}{16 - 0} = \frac{20}{16} = \frac{5}{4}$.

So Linda's plant is growing 1.33 inches per week and Cheryl's plant is growing 1.25 inches per week. Since both rates are expressed in the same units and $1.33 > 1.25$ we conclude that Linda's plant is growing *faster* than Cheryl's.

What happens if we choose different points to calculate the rates? A little experimentation shows that in a linear table it does not matter which two points we use to calculate the rate of change or in which order - we always get the same rate. In fact, this is another property characterizing linear functions:

Definition 2.4 *The Constant Rate of Change*

A **linear function** is a function $y = f(x)$ having a **constant rate of change**. That means that the rates of change between **any** two points (x_1, y_1) and (x_2, y_2) of the function are all the same.

Example 2.1

Refer to **Table 2.4** to answer the following questions:

- Does the table represent a linear function? Explain why or why not.

- IF it DOES represent a linear function, find the constant rate of change (including units) in at least two different ways.

t months	$P(t)$ dollars
3	28,000
7	21,000
11	14,000
15	7,000

Table 2.4

Solution: The successive output differences are constant (all the same) and equal to $-7,000$ dollars as the inputs increase by the constant of 4 months.

$$\Delta t = 7 - 3 = 11 - 7 = 15 - 11 = +4 \text{ months}$$
$$\Delta P = 21,000 - 28,000 = -7,000 \text{ dollars}$$
$$\Delta P = 14,000 - 21,000 = -7,000 \text{ dollars}$$
$$\Delta P = 7,000 - 14,000 = -7,000 \text{ dollars.}$$

Diff	t	$P(t)$	1stDiff
+4	3	28,000	$-7,000$
+4	7	21,000	$-7,000$
+4	11	14,000	$-7,000$
	15	7,000	

Example 2.1 Linear? - Yes

So the table represents a linear function. One way to find the constant rate of change is to take the ratio between the constant first differences:

$$\text{Rate of change} = \frac{\Delta P}{\Delta t} = \frac{-7000 \text{ dollars}}{4 \text{ months}} = \frac{-7000/4 \text{ dollars}}{1 \text{ month}} = -1750 \text{ \$ per month.}$$

Another way to find the constant rate of change is to find the ratio between the last point $(15, 7,000)$ and the first point $(3, 28,000)$ in this order:

$$\text{Rate of change} = \frac{\Delta P}{\Delta t} = \frac{(28,000 - 7,000) \text{ dollars}}{(3 - 15) \text{ months}} = \frac{21,000 \text{ \$}}{-12 \text{ mos}} = -1750 \text{ \$ per month.}$$

(Notice in this table that the output values are decreasing and the constant rate of change is negative.)

Example 2.2

Refer to **Table 2.5** to answer the following questions:

- Does the table represent a linear function? Explain why or why not.

- IF it DOES represent a linear function, find the constant rate of change.

d days	$N(d)$ cells
1	24,000
4	30,000
7	37,500
10	46,875

Table 2.5

Solution:

The output differences are NOT constant as the inputs increase by a constant:

$$(\Delta N)_1 = 30,000 - 24,000 = 6,000 \text{ cells}$$
$$(\Delta N)_2 = 37,500 - 30,000 = 7,500 \text{ cells.}$$

Diff	d	$N(d)$	1stDiff
+3	1	24,000	$+6,000$
+3	4	30,000	$+7,500$
+3	7	37,500	$+9,375$
	10	46,875	

Example 2.2 Linear? - No

No, this table does NOT represent a linear function because the consecutive output values do not change by a common difference although the inputs increase by a constant - there is no common difference between consecutive outputs. Notice also that the rate $(\Delta N)_1/\Delta d = 2,000$ is not equal to the rate $(\Delta N)_2/\Delta d = 2,500$ and the table is not linear regardless of the last point.

Linear Function Equations

Let us reformulate Linda's plant example in terms of rates of change: *Linda bought a seedling plant that is 12 inches tall and it will grow about* **4/3 inches per week** *for the first 15 weeks.* In this context, let $t = 0, 1, 2, 3, \ldots$ weeks and observe the pattern displayed by the height $h(t)$ below:

(Initial Value) $h = h(0) = 12$ $= 12 + (4/3) \cdot \mathbf{0}$

$(+(4/3))$ $h = h(1) = 12 + (4/3)$ $= 12 + (4/3) \cdot \mathbf{1}$

$(+(4/3))$ $h = h(2) = [12 + (4/3)] + (4/3)$ $= 12 + (4/3) \cdot \mathbf{2}$

$(+(4/3))$ $h = h(3) = [12 + (4/3) + (4/3)] + (4/3)$ $= 12 + (4/3) \cdot \mathbf{3}$

                        ...

(Linear Function) $h = h(\mathbf{t}) = [12 + (4/3) + (4/3) + \ldots + (4/3)] + (4/3)$ $= 12 + (4/3) \cdot \mathbf{t}.$

We conclude that the equation of the plant height function is

(Change of order)   **(2.7)** $h(t) = (4/3) \cdot t + 12$ inches, $t = 0, 1, 2, 3, \ldots, 15$ weeks.

Now that we have the equation, we can calculate the height after 15 weeks directly:

(Evaluate) $h(15) = (4/3) \cdot 15 + 12 = 20 + 12 = 32$ inches.

Linear function *equations* (*formulas*) can always be written in the form:

$$y = f(x) = mx + b, \qquad m, b = \text{ constants},$$

> ☞ If $m = 0$, the linear function is constant
>
> $y = mx + b = 0 \cdot x + b = b.$

i.e., are given by *polynomials* where the highest power of the (input) variable is at most 1 (no higher degree terms.) Here m and b also refer to the numerical *coefficients* of the linear (plain 'x') and constant *terms* respectively.

The coefficient m is the *constant rate of change* of y relative to x. The term b is the value of the output y when the input $x = 0$ and is thus the *initial value* of the function:

$$y = f(0) = m \cdot 0 + b = 0 + b = b.$$

Definition 2.5 *Linear Function Equation (Formula)*

A **linear function** can be defined by an equation (formula) of the form

$$y = f(x) = mx + b,$$

where x is the independent variable (input), y is the dependent variable (output), and m, b are constants. Moreover, m represents the **constant rate of change**, and b the **initial value** of the function.

Examples of Linear Functions

The following equations (formulas) define linear functions:

Recall that m goes with the linear term, and b with the constant term.

$$y = x \qquad m = 1, \qquad b = 0$$
$$f(x) = 3 - 2x \qquad m = -2, \qquad b = 3$$
$$h(s) = -40 \qquad m = 0, \qquad b = -40$$
$$y = 4.9t - 5 \qquad m = 4.9, \qquad b = -5.$$

Non-Examples of Linear Functions

The following equations (formulas) do not define linear functions:

$y = 7x^2 - 8$	Not linear because there is a quadratic term. This is a quadratic function.
$V(r) = 1 - x + x^3$	Not linear because there is a cubic term. This is a cubic function.
$y = 1200(1.05)^t$	Not linear because it is not a polynomial function. (The input variable is in the exponent - this is an exponential.)

Negative and Positive Rates of Change

In **Example 2.1** the rate of change has a *negative sign* as the output differences $\Delta P = -7,000$ are negative and the input differences $\Delta t = +4$ are positive:

$$\Delta P / \Delta t = -\$7,000/ + 4 \text{ mos} = -\$1750 \text{ per month.}$$

(Sign rule)

What is the meaning of this sign? Here is a possible scenario matching the table: *An escrow account is opened with a balance of $28,000 to pay a rent of $1,750 per month.* In order to write the correct equation for this linear situation, we identify $m = -1750$, where the negative sign means *payments out* of the account. Since the initial value is $b = 28,000$, the equation for the amount *left* after t months is

$$P(t) = mt + b = -1750t + 28000 \text{ dollars.}$$

Δt	t	$P(t)$	ΔP
$+4$	3	28,000	$-7,000$
$+4$	7	21,000	$-7,000$
$+4$	11	14,000	$-7,000$
	15	7,000	

Example 2.1

Compare this with Linda's plant equation (**2.7**) where the rate of change $m = 4/3$ has a positive sign, which means that the plant is *growing*. We deduce

$m = \Delta y/\Delta x$ is positive if both Δy and Δx have the same sign and negative otherwise.

Fact 2.6 *The Sign of a Rate of Change*

A linear function is **increasing** when the rate of change is **positive**, $m > 0$, **decreasing** when the rate of change is **negative**, $m < 0$, and **constant** when the rate of change is **zero**, $m = 0$.

Examples of Linear Function Equations in Context

For each example below, 1) identify and label the independent & dependent variables, the initial value (b), and the constant rate of change (m) (all including units), and 2) write an equation of a linear function summarizing the situation.

☞ Time is usually chosen as the independent variable.

a) Linda bought a plant that is two feet tall and it will grow about 1/4 inch per week.

> IV (Input): t = the time since the plant was bought (**unit:** weeks)
> DV (Output): $h(t)$ = the height of the plant (**unit:** inches)

The initial value is $b = 2$ feet $= 24$ inches; the constant rate of change is $m = 1/4$ inch per week; the equation is $h(t) = mt + b = (1/4)t + 24$ inches.

☞ Include units at each stage and convert them as needed.

b) A car rental company charges $25 and an additional 2 cents for every mile driven.

> IV (Input): d = the distance traveled by the car (**unit:** miles)
> DV (Output): $C(d)$ = the total cost of the rental (**unit:** dollars)

The initial value is $b = 25$ dollars; the constant rate of change is $m = 2$ cents per mile $= 0.02$ dollars per mile; the equation is $C(d) = md + b = 0.02d + 25$ dollars.

☞ The constant rate in a linear situation is the change in output per change in input and thus its numerator is the output variable and its denominator the input.

c) The pool started with 14,000 gallons in it and drained 12 gallons every 3 minutes.

> IV (Input): t = the time since the pool is drained (**unit:** minutes)
> DV (Output): $W(t)$ = the amount of water left in the pool (**unit:** gallons)

The initial value is $b = 14,000$ gallons. To find the constant rate of change we take the ratio of the change in W to the change in t:

$$m = \frac{\Delta W}{\Delta t} = \frac{-12 \text{ gallons}}{3 \text{ minutes}} = \frac{-12/3 \text{ gallons}}{1 \text{ minute}} = -4 \text{ gallons per minute.}$$

The equation is $W(t) = mt + b = -4t + 14,000$ gallons.

Linear Model properties

Constant differences in output if inputs increase by a constant.

Constant rate of change:

$$\frac{\text{Diff } y}{\text{Diff } x} = \frac{\Delta y}{\Delta x} = \frac{y_2 - y_1}{x_2 - x_1}.$$

Graphs are (straight) lines.

Function equations $y = mx + b$ with rate m and initial value b.

Table 2.6

Wrapping Up

Functions describe (model) how two variables are related and they help us to answer questions about real-world situations. By modeling plant-growing situations in this section we discovered properties of a specific type of model - linear functions. See **Table 2.6**. The points of a linear function 'line up' on the graph, which justifies its name. The linear function tables are characterized by constant output differences when inputs increase by a constant. They are also characterized by constant rates of change - the ratios $\Delta y / \Delta x$ are always the same no matter which two points we choose or in which order. Increasing linear functions have positive rates of change and decreasing linear functions have negative rates of change. In the next section we will learn more about the graphs of linear functions and we will find equations for linear models from tables, graphs, and situations.

Exercises

Exercises for 2.1 Introduction to Linear Relationships

P2.1 Do you think that plants grow linearly over their entire life time? Explain.

P2.2 For each table below do the following tasks:

- Determine whether or not the table represents a linear function, and *show how you know*.

- For tables that are linear, find the constant rate of change with different sets of points.

a)

n books sold	c dollars raised
0	250
2	264
4	278
6	292
8	306

b)

n toppings	c pizza cost in $
0	7.50
2	8.00
4	8.50
6	9.00
8	10.50

c)

d days	$N(d)$ cells
1	45,000
3	30,000
5	20,000
7	13,333

d)

t reading hrs.	p pages left to read
3	275
5	175
5.5	150
7	75

P2.3 For each function below do the following tasks:

- Determine whether it is linear or not and explain why.

- If it is linear, identify the constant rate of change and the initial value.

a) $y = 43x + 7$

b) $f(x) = 3 - 2x$

c) $h(s) = -40$

d) $f(x) = 2(.06)^x$

e) $y = 7x^2 - 8$

f) $y = x$

g) $r(t) = 64t$

h) $k(x) = 32x - 4$

i) $V(r) = 1 - r + r^3$

P2.4 For each linear situation below,

- Identify and label the variables and the constants including units.

- Identify the constant rate of change and the initial value.

- Write an equation for the linear function that relates the two variables.

a) The pool started with 14,000 gallons in it and then drained 12 gallons every minute.

b) The diver rose towards the surface of the water 30 feet each minute from 390 feet *below* the surface.

c) The pool started with 30,000 gallons of water in it and then drained 12 gallons every 3 minutes.

d) There are 30 grams of sugar in every 50 grams of milk chocolate M&Ms.

e) A baby grows about 1/2 inch each month for the first year. She was born a foot and a half long.

f) A fitness club charges $240 for the year and $3 for each class.

g) Mel makes monthly car payments of $420 which are automatically withdrawn from a bank account. She originally deposited $15,000 into that account and no other transactions will take place on that account which does not earn interest.

h) A plane starts to descend 400 feet every minute after it was cruising at 10,400 feet.

i) Padma's new business started with $23,000 of debt (or negative profit) and then the business lost another $500 a week for the first year.

j) The flood waters in Springfield were 5 feet then every day the water receded 6 inches.

k) A submarine 600 feet below the surface comes up 2.5 feet per second.

l) Sweet Stuff Berry farm has pick-your own berries. They charge $3 for each quart picked.

m) The temperature control system in an office maintains the temperature at 68° Fahrenheit all day long.

P2.5 Verify that the function table is linear by checking the rate of change between different pairs of points to see that they are all the same - that there is a constant rate of change. How many pairs do you need to verify? If it is linear, write the constant rate using correct units.

a)

months	inches
1	7
2	11
5	23

b)

toppings	total cost in $
4	0
6	3
10	9
12	12

c)

days	cells
4	200
5	600
7	1800

d)

time s	velocity m/s
0	0
2	−4
3	−6
4	−8

2.2 Analyzing Linear Functions ($y = mx + b$)

There are many situations in the real world with constant rates of change between two variables and to analyze them we use linear models in the forms of graphs, tables, and equations. They help us to better understand such situations and to answer questions about them. In this section we will learn how to create linear models and how to interpret their key features - slope and intercepts.

Launch Exploration

For the graphs in **Fig 2.5** below answer the following questions:

Can you write the equation in the form $y = mx + b$?
    IF yes, identify the **coefficients** m and b.
Identify the **increasing** graphs and the **decreasing** graphs. What do the equations for the increasing graphs have in common? The decreasing graphs?
Do all the graphs have a y-**intercept**? Explain. Are there any x-**intercepts**?
    If yes, how many?
Which graph is a graph of a function which is **neither increasing nor decreasing**?
What kind of function is it? Which graph is not the graph of a **function**? Explain.

Estimate the coordinates (x, y) of all important points and fill in the **Table 2.7**.

$y = mx + b$		
Point	**(x, y)**	**Coeff.**
y- intercept	?	$b = ?$
x- intercept	?	

Table 2.7 Important Points

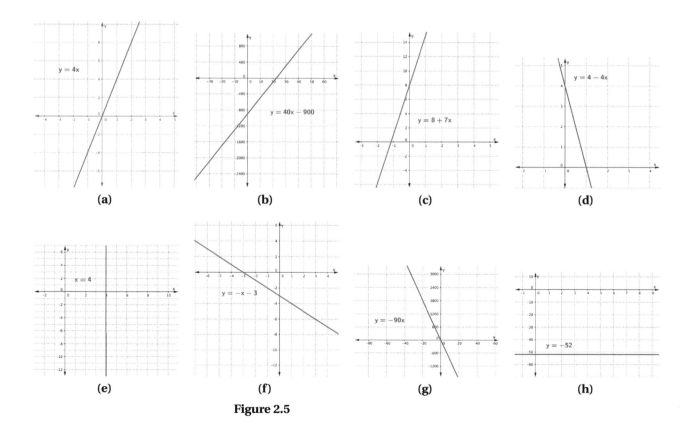

(a) (b) (c) (d)

(e) (f) (g) (h)

Figure 2.5

Exploring Linear Graphs

By inspection, all equations in the **Launch Exploration** can be written in the form $y = mx + b$ except for the equation $x = 4$ of the vertical line in **Fig 2.5 (e)**. Notice that in **Fig 2.5 (d)**, $4 - 4x = -4x + 4$ so the *linear* coefficient is $m = -4$ and in **Fig 2.5 (c)**, $8 + 7x = 7x + 8$ so that $m = 7$. Here is a summary of our graph analysis:

Equation	m	b	Direction	y-Intercept	x-Intercept
$y = 4x + 0$	4	0	increasing	$(0, 0)$	$(0, 0)$
$y = 40x - 900$	40	-900	increasing	$(0, -900)$	$(22.5, 0)$
$y = 8 + 7x$	7	8	increasing	$(0, 8)$	$(-\frac{8}{7}, 0)$
$y = 4 - 4x$	-4	4	decreasing	$(0, 4)$	$(1, 0)$
$x = 4$	undefined	?	vertical	none	$(4, 0)$
$y = -x - 3$	-1	-3	decreasing	$(0, -3)$	$(-3, 0)$
$y = -90x + 0$	-90	0	decreasing	$(0, 0)$	$(0, 0)$
$y = 0x - 52$	0	-52	horizontal	$(0, -52)$	none

The increasing graphs go up from left to right and their equations have the coefficient m *positive*. The decreasing graphs go down from left to right and their equations have the coefficient m *negative*. The *constant* function $y = -52$ is neither increasing nor decreasing and has the coefficient $m = 0$. The vertical line is not the graph of a function since the same input, $x = 4$, has infinitely many outputs. In this case, the coefficient m is *undefined*.

Fact 2.7 *Linear Graph Direction*

The graph of a linear function $y = mx + b$ is a **(straight) line**. When traced from left to right, the line goes up (**increasing**) when $m > 0$, goes down (**decreasing**) when $m < 0$, and is horizontal (**constant**) when $m = 0$. Equations of the form $x = a$ are **vertical** lines for which m in undefined.

Recall that a *y-intercept* is a point where a graph crosses the y-axis and thus must have a first coordinate 0. An *x-intercept* is a point where a graph crosses the x-axis and thus must have a second coordinate 0. All the graphs in the **Launch Exploration** have one *y-intercept* except for the vertical line $x = 4$ which has no y-intercepts. All the graphs have one *x-intercept* except for the horizontal line $y = -52$ which has no x-intercepts.

☞ The y-axis is a vertical line with *infinitely many y-intercepts*.

The x-axis is a horizontal line with *infinitely many x-intercepts*.

Fact 2.8 *Linear Graph Intercepts*

The y-**intercept** of a linear graph $y = mx + b$ is the **point** $(0, b)$ having **zero** as its x-coordinate, $x = 0$, and the **initial value** as its y-coordinate, $y = b$. The x-**intercept** is the point $(x, 0)$ having **zero** as its y-coordinate, $y = 0$, and the **solution** to the equation $0 = mx + b$ as its x-coordinate.

Finding the y-intercept given a linear equation

To find the *exact* y-intercept of each graph in **Fig 2.5**, we set $x = 0$ and evaluate y:

(b)	$y = 40x - 900$	**(c)**	$y = 8 + 7x$	**(d)**	$y = 4 - 4x$
	$y = (40)(0) - 900 = -900$		$y = 8 + (7)(0) = 8$		$y = 4 - (4)(0) = 4$
	$(0, -900)$		$(0, 8)$		$(0, 4)$
(e)	$x = 4$	**(g)**	$y = -90x$	**(h)**	$y = -52$
	$0 = 4$		$y = (-90)(0) = 0$		$y = -52$
	no y-intercept		$(0, 0)$		$(0, -52)$

Notice that the equation $y = -52$ holds for *any* x-value including $x = 0$, so the y-intercept is $(0, -52)$. Notice that the equation $x = 4$ holds for any y-value. However when we substitute $x = 0$ into the equation, we get $0 = 4$ which is a contradiction, so the line has no y-intercept.

Finding the x-intercept given a linear equation

To find the *exact* x-intercept of each graph in **Fig 2.5**, we set $y = 0$ and solve for x:

(b)	$y = 40x - 900$	**(c)**	$y = 8 + 7x$	**(d)**	$y = 4 - 4x$
	$0 = 40x - 900$		$0 = 8 + 7x$		$0 = 4 - 4x$
	$x = 900/40 = 22.5$		$x = -8/7$		$x = 1$
	$(22.5, 0)$		$(-8/7, 0)$		$(1, 0)$
(e)	$x = 4$	**(g)**	$y = -90x$	**(h)**	$y = -52$
	$y = 0$		$0 = -90x$		$0 = -52$
	$x = 4$		$x = 0$		no solution
	$(4, 0)$		$(0, 0)$		no x-intercept

Notice that the equation $x = 4$ holds for *any* y-value including $y = 0$, so the x-intercept is $(4, 0)$. Notice that the equation $y = -52$ holds for any x-value. When we substitute $y = 0$ into the equation, we get $0 = -52$ which is a contradiction, so the line has no x-intercept.

Graphing linear equations

Sometimes we have only the equation of a function and we need to draw its graph.

a) Graph the equation $y = 8 + 7x$.

We know the graph is increasing because $m = 7$ is positive. Next we need two points to determine the line and a third point to verify it. We first find the intercepts $(0, 8)$ and $(-8/7, 0)$ as above - they are often important in modeling. Let the test point be $x = 1$, $y = 8 + 7(1) = 15$. Enter the ordered pair solutions in a **Table 2.8**, plot them, connect them by a solid line, and get the graph **Fig 2.5** (**c**).

☞ An intercept is a point $(a, 0)$ or $(0, b)$ where the graph crosses an axis. Sometimes the number a is called an x-intercept and b a y-intercept.

☞ The equation of a vertical line is of the form $x = a$.

The equation of a horizontal line is of the form $y = b$.

Input x	Output y	
0	8	y-intercept
$-8/7$	0	x-intercept
1	15	extra pt

Table 2.8 Points to graph **Fig 2.5** (**c**) $y = 8 + 7x$.

Input x	Output y	
0	0	x-int & y-int
10	−900	extra pt
−20	1800	extra pt

Table 2.9 Points to graph **Fig 2.5** (**g**) $y = -90x$.

b) Graph the equation $y = -90x$.

Since $m = -90$ is negative, the graph is decreasing. Both intercepts are at $(0, 0)$ and you need two additional points. For example, $x = 10$, $y = -90(10) = -900$ and $x = -20$, $y = -90(-20) = 1800$. After you enter them in **Table 2.9**, plot and connect them by a solid line, you should get the graph **Fig 2.5** (**g**).

Writing Linear Function Equations from Graphs

Given the graph of a linear function, to write its equation $y = mx + b$, we need to find the two coefficients m and b. In geometry, the coefficient m is called *slope*.

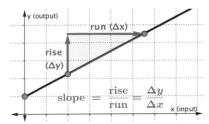

Figure 2.6 Slope Triangle

Definition 2.9   *The Slope of a Line*

The **slope** m of a line is the ratio of **rise** over **run** between *any* two points (x_1, y_1) and (x_2, y_2) of the line:

$$\text{slope} = \frac{\textbf{rise}}{\textbf{run}} = \frac{\textbf{change in } y}{\textbf{change in } x} = \frac{\textbf{Diff } y}{\textbf{Diff } x} = \frac{\Delta y}{\Delta x} = \frac{y_2 - y_1}{x_2 - x_1} = m.$$

The **slope** m of a line also represents the **(constant) rate of change** of a linear function whose graph is that line.

The **rise** from one point to another is the difference Δy between their y-coordinates (vertical change) and the **run** is the difference Δx between their x-coordinates (horizontal change). To find the slope of a line, we can draw a right triangle as in **Fig 2.6** that shows the vertical and horizontal changes (rise and run) between two points on the line. The ratio of rise over run will be the same for any two points on the line as all such "slope triangles" are similar.

For example, a keg is drained according to the graph in **Fig 2.7**.

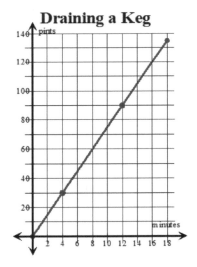

Figure 2.7

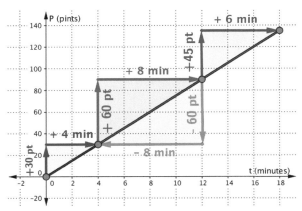

Figure 2.8 Draining a Keg Analysis

The independent variable is t - the draining time in minutes, the dependent variable is P - the amount of fluid taken out from the keg. The slope is constant as all triangles in the diagram **Fig 2.8** are similar:

$$m = \text{slope} = \frac{\text{rise}}{\text{run}} = \frac{+30\text{ pt}}{+4\text{ min}} = \frac{+60\text{ pt}}{+8\text{ min}} = \frac{-60\text{ pt}}{-8\text{ min}} = \frac{+45\text{ pt}}{+6\text{ min}} = 7.5\text{ pt/min}.$$

run Δt	t	P	rise ΔP
+4	0	0	+30
+8	4	30	+60
+6	12	90	+45
	18	135	

Table 2.10 Keg-draining Table

The slope tells us that the fluid is removed at a constant rate of 7.5 pints per minute. The y-intercept is $(0,0)$ which means that the initial value of the function is 0 pints removed after 0 minutes of draining. See **Table 2.10**.

The equation of a line which is not vertical can be written in the form

(**2.8**)
$$y = mx + b$$
(Slope-Intercept Form)

where m is the slope (rate of change) and b is the y-intercept (initial value). In the keg-draining case, $m = 7.5$, $b = 0$, and the equation is $P = 7.5t$ or in function notation $P(t) = 7.5t$.

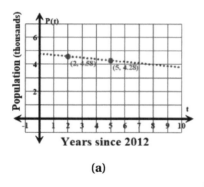
(a)

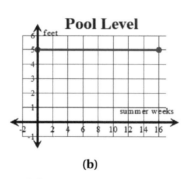
(b)

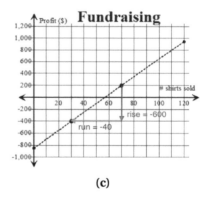
(c)

Figure 2.9

Examples of writing equations from graphs

For each graph above do the following tasks:

- Identify and label the independent and dependent variables.
- Find the slope and explain what it means.
- Identify or find the initial value.
- Write an equation for the linear function.
- Use your equation to find the exact 'x'- and 'y'-intercepts.

a) The independent variable is the time t in years since 2012. The dependent variable is the number of people P in thousands. To find the slope we use two data points from the graph, $(2, 4.58)$ and $(5, 4.28)$. See **Table 2.11**.

run Δt	t	$P(t)$	rise ΔP
+3	2	4.58	−0.3
	5	4.28	

Table 2.11

$$m = \frac{\Delta P}{\Delta t} = \frac{4.28 - 4.58}{5 - 2} = \frac{-0.3 \text{ thous.}}{+3 \text{ years}} = -0.1 \text{ thous. people per year.}$$

The minus sign means the population *decreases* at a rate of 100 people per year. The equation of the line is now $P = -0.1t + b$ and we still need to find the y-intercept b. We can get an estimate from the graph, but to find the *exact* value recall that any point on the graph is an *ordered pair solution* to the equation. For example we can choose $(t, P) = (2, 4.58)$ and if we substitute these values into the equation $P = -0.1t + b$, we get an equation that lets us *solve* for b!

$$4.58 = -0.1(2) + b, \qquad b = 4.78, \qquad P(t) = -0.1t + 4.78.$$

The P-intercept is $P(0) = -0.1(0) + 4.78 = 4.78$ thous. people or $(0 \text{ years}, 4.78 \text{ thous.})$ and represents the initial population in 2012. The t-intercept is obtained by setting $P = 0$ in the equation and solving for t:

$$0 = -0.1t + 4.78, \qquad -4.78 = -0.1t, \qquad t = 47.8 \text{ years.}$$

The population becomes extinct after 47.8 years according to this model.

b) The dependent variable is the water depth D in feet. The independent variable is the time t in weeks. For slope we use the points $(0, 5)$ and $(6, 5)$. See **Table 2.12**.

$$m = \frac{\Delta D}{\Delta t} = \frac{5 - 5}{6 - 0} = \frac{+0 \text{ ft}}{+6 \text{ weeks}} = 0 \text{ feet per week.}$$

Hence, the function is the constant function $D(t) = 5$. There are no t-intercepts. This model shows a constant water level in a pool during summer.

c) The independent variable is the number s of shirts sold. The dependent variable is the amount of profit P in dollars. For the slope we use the marked lattice points $(30, -400)$ and $(70, 200)$ on the graph. See **Table 2.13**.

$$m = \frac{\Delta P}{\Delta t} = \frac{200 - (-400)}{70 - 30} = \frac{+\$600}{+40 \text{ shirts}} = \frac{-\$600}{-40 \text{ shirts}} = 15 \text{ \$ per shirt.}$$

If we substitute $m = 15$ and $(s, P) = (30, -400)$ into the equation $P = ms + b$, we get

$$-400 = (15)(30) + b, \qquad b = -850, \qquad P(s) = 15s - 850.$$

The P-intercept is $P(0) = 15(0) - 850 = -\$850$ or $(0 \text{ shirts }, -\$850)$ and is the amount of money *paid* before selling. The t-intercept is the solution to the equation

$$0 = 15s - 850, \qquad 850 = 15s, \qquad s = 850/15 \approx 57 \text{ shirts.}$$

To break even, zero loss and zero profit, the club must sell about 57 shirts.

0.1 thousand people
= 1 tenth of a thous. people
= $\frac{1}{10}(1000)$ people
= 100 people

run Δt	t	$D(t)$	rise ΔD
+6	0	5	+0
+6	6	5	+0
+6	12	5	+0
	18	5	

Table 2.12

run Δs	s	$P(s)$	rise ΔP
+40	30	−400	+600
	70	200	

Table 2.13

Writing Linear Function Equations from Tables

When we write linear equations from graphs we look for two data points on a graph that have exact coordinates that are easy to find, so we can calculate the exact slope. To avoid mistakes in calculations, it can be best to enter the points from the graph into a table. Then find the rise Δy and the run Δx between two points by going in the SAME direction on the table like we did in **Section 2.1**. The ratio will be the slope and also the constant rate of change. Then we substitute a point which is on the line into the equation $y = mx + b$ to find the y-intercept.

For example, when we wrote the equations for the graphs (**a**), (**b**), (**c**) in **Fig 2.9** we used the **Tables 2.11, 2.12**, and **2.13**. Here are a few more examples of tables:

t = Amount of Time a Candle Burns (hours)	H(t) = Height of the Candle (cm.)
0	24
2	21
4	18
8	12

(**a**)

t = Time Traveled (hrs.)	D(t) = Distance Driven (km.)
2	220
3	320
4	420
7.5	770

(**b**)

t = hours since 5pm	H(t) = Height of the Candle (inches)
0.5	8
1	8
5	8
14	8

(**c**)

Figure 2.10 Tables

Examples of writing equations from tables

For each table above do the following tasks:

- Identify and label the independent and dependent variables.
- Determine if the function table is linear by checking the rates of change to see if they are the same.

If the table is linear then

- Find the constant rate of change (slope) including units.
- Identify or find the initial value (y-intercept) including units.
- Write an equation for the linear function.

☞ The term 'y-intercept' is used for both the point $(0, b)$ and the initial value b.

a) The independent variable is the burning time t in hours. The dependent variable is the candle height H in cm. To determine whether the table is linear, we calculate the rate of change between successive points in **Table 2.14**.

run Δt	t	$H(t)$	rise ΔH
+2	0	24	−3
+2	2	21	−3
+4	4	18	−6
	8	12	

Table 2.14

$$m = \frac{\Delta H}{\Delta t} = \frac{-3 \text{ cm}}{+2 \text{ hrs}} = \frac{-6 \text{ cm}}{+4 \text{ hrs}} = -1.5 \text{ cm per hour.}$$

Since all successive rates of change are equal, the table is linear and $m = -1.5$ is the slope. The negative sign indicates that the candle *melts* at a rate of 1.5 cm per hour. To write the equation of the function, we can read the initial value directly from the table, $H(0) = 24$ cm. So the equation is $H(t) = -1.5t + 24$.

run Δt	t	$D(t)$	rise ΔD
+1	2	220	+100
+1	3	320	+100
+3.5	4	420	+350
	7.5	770	

Table 2.15

b) The independent variable is the time traveled t in hrs. The dependent variable is the distance traveled D in km. The successive rates in **Table 2.15** are all the same:

run Δt	t	$D(t)$	rise ΔD
-3.5	7.5	770	-350
-1	4	420	-100
-1	3	320	-100
-1	2	220	-100
-1	1	120	-100
	0	20	

Table 2.16 Reversed Table

$$m = \frac{\Delta D}{\Delta t} = \frac{+100 \text{ km}}{+1 \text{ hr}} = \frac{+350 \text{ km}}{+3.5 \text{ hrs}} = 100 \text{ km per hour.}$$

Hence, the table is linear with the constant slope $m = 100$ km per hour (speed). As the initial value is not listed in the table, it is necessary to work backwards, which means that it is necessary to SUBTRACT the difference instead of adding it. More specifically, we go backwards in the table from the point $(2, 220)$ to the point $(0, b)$ by successively subtracting the difference 100 from the output value two times:

$$(2, 220) \rightarrow (1, 120) \rightarrow (0, 20) \qquad b = 20 \text{ km.}$$

Another way to view this procedure is to write a reversed **Table 2.16** in the margin to eventually arrive at the initial output when the initial input $t = 0$. The equation of the function $D = mt + b$ is $D(t) = 100t + 20$.

c) The independent variable is the time t in hours. The dependent variable is the candle height H in inches. To determine whether the table is linear, we calculate the rate of change between successive points in **Table 2.17**.

run Δt	t	$H(t)$	rise ΔH
$+0.5$	0.5	8	$+0$
$+4$	1	8	$+0$
$+9$	5	8	$+0$
	14	8	

Table 2.17

$$m = \frac{\Delta H}{\Delta t} = \frac{+0 \text{ in}}{+0.5 \text{ hrs}} = \frac{+0 \text{ in}}{+4 \text{ hrs}} = \frac{+0 \text{ in}}{+9 \text{ hrs}} = 0 \text{ in per hour.}$$

Since all successive rates of change are equal, the table is linear and $m = 0$ is the slope. This means that the function is constant and thus we can write the equation $H(t) = 8$. The candle is not burning.

Comparing Slopes

Remember Linda's and Cheryl's plants from **Section 2.1**? Their graphs are plotted in the margin, **Fig 2.11**. Notice that Linda's graph is *steeper* than Cheryl's graph. What does that mean? The slope of a line tells us the direction of the line (up if positive or down if negative) and how steep it is: the greater the *absolute value* of the slope the steeper the line. The slope is also the rate of change, so a steeper line translates into a greater, faster rate of change. Indeed, we have seen in **Section 2.1** that Linda's plant grows at a rate of **4 in/3 wk** ≈ 1.33 inches per week which is faster than **5 in/4 wk** $= 1.25$ inches per week, the growth rate for Cheryl's plant.

In order to compare slopes by looking at graphs, it is crucial to have these graphs plotted on the SAME scale. For example, comparing the two keg-draining graphs in **Fig 2.7** and **Fig 2.8**, the first graph looks steeper than the second, but the two graphs are identical if we plot them on the same scale.

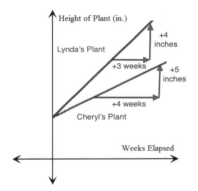

Figure 2.11 Which plant grows faster?

Example 2.3 For each diagram in **Fig 2.12** below, answer the following questions and EXPLAIN your reasoning.

- Which car is the fastest?
- Which car is the slowest?

- Which car is parked?

Solution: In each diagram we can estimate the slopes by making slope triangles as suggested by the dotted lines shown in the diagrams and then doing a rise over run analysis including units. These slopes represent the velocities of the cars, i.e. the speed and direction. A positive slope means that the distance from home is increasing and thus the car drives away from home and a negative slope means the distance from home is decreasing and thus the car drives towards home. A zero slope means that the distance from home is constant (stays the same).

☞ speed = $\frac{\text{distance traveled}}{\text{time elapsed}}$

Diagram (a) Speed A is zero mph (horizontal line has slope zero), speed B is **80/2** = 40 mph, and speed C is **50/2** = 25 mph both away from home. B is the fastest, C is the slowest, and A is parked.

Diagram (b) Speed A is **50/2** = 25 mph towards home, speed B and speed C are both **50/2** = 25 mph away from home. All cars have the same speed.

Diagram (c) Speed A is **30/1** = 30 mph, speed B is **30/2** = 15 mph, and speed C is **50/2** = 25 mph all away from home. A is the fastest and B is the slowest.

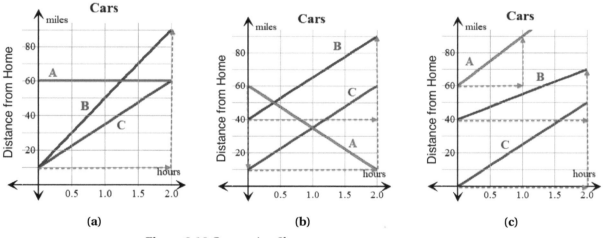

Figure 2.12 Comparing Slopes

Fact 2.10 *Parallel Lines*

Two lines are **parallel** if and only if they have the **same slope (rate of change)** but different intercepts.

Wrapping Up

A linear relationship between an independent variable (input) and a dependent variable (output) can be determined by just two characteristics: its initial value, and its constant rate of change. With those two we can write a linear equation

(**2.9**) $$y = mx + b$$

where x is the input and y is the output.

What is m?
- m is the constant rate of change
- m is the slope of the graph

(**2.10**) $$m = \frac{\Delta y}{\Delta x} = \frac{\text{Rise}}{\text{Run}} = \frac{\text{Change in Outputs}}{\text{Change in Inputs}}$$

How to find m?
- In a verbal description look for a constant rate between two variables (miles *per* hour, calories *in every* serving, miles *per* gallon, etc.)
- In a linear table, calculate the ratio between the change in two outputs to the change in their inputs.
- In a linear graph, make a slope triangle to estimate the rise and the run and take the ratio of the rise over the run.
- In a linear equation, m is the coefficient of the input variable.

What does m mean?
- m shows the steepness of a linear graph or how fast the output is changing.
- A positive slope means the linear function and its graph is increasing.
- A negative slope means the linear function and its graph is decreasing.
- A zero slope means a horizontal line (no change in output).
- An undefined slope means a vertical line (unlimited change in output).

What is b?
- b is the initial value of the function (when the input is 0).
- b or rather the point $(0, b)$ is the y-intercept of the graph.

How to find b?
- In a verbal description look for the starting, initial value of the output variable.
- In a linear table, look for the entry with input of 0. If it is not in the table, you can go backwards in the table using the constant differences or substitute a known point into the equation and solve for b.
- In a linear equation b is the constant term.

In the next section we show how to create linear models for various applications and how to use them to make predictions.

Exercises

Exercises for 2.2 Analyzing Linear Functions ($y = mx + b$)

P2.6   For each graph below do the following tasks:

- Identify and label the independent and dependent variables.
- Find the slope and explain its meaning in context including units.
- Find the 'y'-intercept and explain its meaning.
- Write an equation for the line.
- Use the equation to find the 'x'-intercept, if possible, and interpret its meaning or explain why it has no meaning in the real-world situation the graph describes.

a) The graph shows the number of gallons of gas left in a tank after driving a certain number of miles.

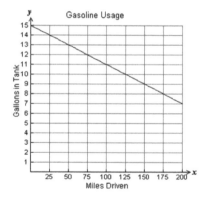

b) The graph shows the weekly wages an employee can earn at Company X.

c) The graph shows the distance traveled by train after a certain number of hours.

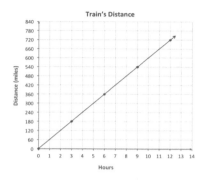

d) The graph shows the number of cricket chirps as a function of ambient temperature.

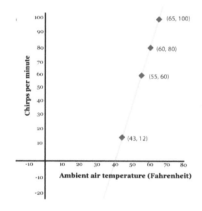

e) The graph shows the number of books read over the summer break.

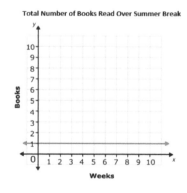

f) The graph shows the number of inches of flood water in a Louisiana town after a major storm.

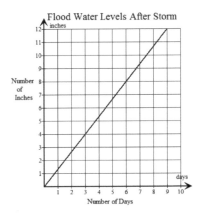

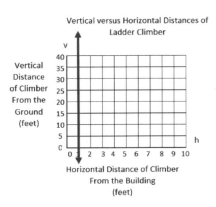

g) The graph shows the vertical distance of a climber in terms of its horizontal distance.

P2.7   For each graph below, write an equation that could represent the line. You don't need to label scales BUT you can earn extra credit if you add scales that match your equation.

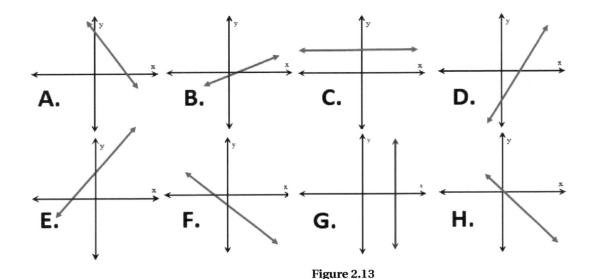

Figure 2.13

Exercises for 1.4 Relationships between Variables - Introduction to Functions

P2.8   For each linear situation:

- Identify the independent & dependent variables. Are they continuous or discrete?

- Explain how to tell whether the relationship between the variables is linear or not. Find the slope and explain its meaning in context including units.

- Find the 'x'- and the 'y'-intercepts and explain their meanings including units or why they do not have a meaning in the given context. Start a table and enter the intercepts.

- Find at least 3 additional ordered pairs, express them in function notation, and enter them in the table. Give a real-world domain and a range for the model.

- Draw and label axes with variables and units, use appropriate scales, and graph the function. Is it increasing or decreasing? Does your answer agree with your equation? Explain.

a) Jerome is saving money by making monthly deposits into a savings account until he has exactly $6,100 to buy a car. This situation is described by the function $A(m) = 100 + 120m$, where m is the number of months and $A(m)$ is the total amount of money, in dollars.

b) At the Working-It-Out gym, the total cost in dollars of an annual membership up to 100 sessions is given by the function equation $C(n) = 180 + 16n$, where n is the number of personal training sessions.

c) The trip of a motor boat is modeled by the function $D(h) = 178 - 12h$, where h is the number of hours since noon and $D(h)$ is the distance from the boat to the marina.

d) In a 6-hour shift a barista can serve up to 270 customers. Her total pay in dollars for one shift is modeled by the equation $P(n) = 0.45n + 66$ where n is the number of customers she serves.

e) Keely goes skiing from the top of a mountain trail and the function $E(s) = 2700 - 45s$ gives her elevation, $E(s)$ in feet, after s seconds of skiing.

f) A fishing boat is 92 miles away from the marina, traveling directly towards it at 8 miles per hour. The captain wants to keep track of how far the boat is from the marina as they travel.

g) A barista at the campus coffee shop makes an average of 50 cents from each customer. At noon one day he counts $18.50 in his tip jar. He has 4 hours left on his shift and could serve up to 160 customers in that time.

h) Tyrell goes skiing. He starts at the top of the mountain which has an elevation of 3000 feet and he skis down the mountain at an (average) speed of 45 miles an hour.

i) At a fair each ride costs $2.25, and Kevin gives his little sister Julia $36 to spend on rides. Julia wants to keep track of how much money she has left while at the fair.

P2.9 For each of table below determine whether or not it is linear. If yes, write a linear function equation that models the data. Interpret the slope as a rate of change using the correct units.

a)

n toppings	c pizza cost in $
0	7.50
2	8.00
4	8.50
6	9.00
8	10.50

b)

d days	$N(d)$ cells
1	45,000
3	30,000
5	20,000
7	13,333

c)

t reading hrs.	p pages left to read
3	275
5	175
5.5	150
7	75

d)

months	inches
1	7
2	11
5	23

e)

toppings	total cost in $
4	0
6	3
10	9
12	12

f)

days	cells
4	200
5	600
7	1800

g)

time s	velocity m/s
0	0
2	−4
3	−6
4	−8

2.3   Creating and Using Linear Models ($y = mx + b$)

Launch Exploration

Window washers have just finished and they begin to hoist the platform down the side of a building. The function $h(t) = 168 - \frac{1}{2}t$ gives the height in feet off of the ground of the pulley-bucket (platform) t seconds after they began to hoist it down.

Which is the independent and which is the dependent variable in this situation? Use the function equation to answer the following questions. Express each answer as an ordered pair and in function notation. Enter each of your answers in a table.

Figure 2.14

a) How tall is the building?

b) How high off the ground is the platform after one minute?

c) How long will it take them to reach the ground?

d) What are the domain and range of the model? Explain.

Creating and Using Linear Models

When we are given an equation to model a linear situation, we can use it to answer questions and make predictions.

IV (In)	DV (Out)
t (sec)	*h*(*t*) (ft)
0	168
60	138
336	0

Table 2.18 Ordered pair solutions.

Window Washers

By inspecting the function equation $h(t) = 168 - \frac{1}{2}t$ and translating its verbal description, we identify

IV (Input): t = the descending time (**unit:** seconds)
DV (Output): $h(t)$ = the height of the platform off the ground (**unit:** feet)

a) To find out how tall the building is we notice that at time $t = 0$ seconds the platform is $h(0) = 168 - \frac{1}{2}(0) = 168$ feet above the ground and based on this information the building should be at least 168 feet tall.

Answer: (**0 sec, 168 ft**), $h(0) = 168$ ft.

b) After 1 minute the height of the platform is given by the value of the function at $t = 60$ seconds, since the variable t is in seconds and we must convert 1 minute into 60 seconds. So the platform is $h(60) = 168 - \frac{1}{2}(60) = 138$ feet off the ground.

Answer: (**60 sec, 138 ft**), $h(60) = 138$ ft.

c) To reach the ground, the height of the platform must be zero feet above the ground. So to find the time we set $h(t) = 0$ in the equation and solve for t:

$$0 = 168 - \tfrac{1}{2}t, \qquad -168 = -\tfrac{1}{2}t, \qquad t = 336 \text{ seconds.}$$

Answer: (**336 sec, 0 ft**), $h(336) = 0$ ft.

d) The time is a continuous variable that starts at $t = 0$ seconds and ends at $t = 336$ seconds when the platform reaches the ground and stops moving.

Real-world domain $=$ The interval of all reals from 0 to 336 sec $=$ **[0, 336] sec.**

Real-world range $=$ The interval of all reals from 0 to 168 ft $=$ **[0, 168] ft.**

When a linear situation is given verbally, we first find the equation and then use it to answer questions. Here is an example.

One-Day Repair Jobs

For one-day repair jobs by Sam there is a linear relationship between the amount of time the repair takes and the total cost of the job. Sasha paid Sam $180 for a repair that took 90 minutes. Paolo paid Sam $240 for a repair that took 3 hours.

a) *Suppose we want to know how much a 2-hour repair would cost. Do you have enough information to answer the question? Explain.*

If we know that a situation is linear, we only need two data points to write an equation from which we can find any other ordered pair. In this case we are given the data points (90 min, $180) which converts to (1.5 hrs, $180) and (3 hrs, $240). So we will be able to write an equation.

b) *What does it mean when we say that there is a linear relationship between two quantities?*

It means that the ratio between the change in output values to the corresponding change in input values is constant.

c) *Which quantity is the independent variable and which is the dependent variable in this situation?*

The cost of a repair job depends on the time of repair. Hence, we choose

| | IV (Input): | $t = $ the repair time (unit: hrs) |
| | DV (Output): | $C(t) = $ the cost of repair job (unit: $) |

Δt	t	$C(t)$	$\Delta C(t)$
+1.5	1.5	180	> +60
	3	240	
	2	200	
	7	400	
	8	440	
	0	120	

Table 2.19 Ordered pair solutions for Sam's repair job.

d) *Write a linear equation relating the two variables.*

The linear equation is of the form $C(t) = mt + b$ where m is the constant rate of change (slope) and b is the initial value. To find m we use the two data points (1.5 hrs, $180) and (3 hrs, $240):

$$m = \frac{\text{Change in output}}{\text{Change in input}} = \frac{\$240 - \$180}{3 \text{ hrs} - 1.5 \text{ hrs}} = \frac{+\$60}{+1.5 \text{ hrs}} = \$40 \text{ per hour.}$$

To find b we substitute the point (3 hrs, $240) and the constant rate $m = \$40$ per hour into the equation and solve for b:

$$C(t) = mt + b, \qquad 240 = (40)(3) + b, \qquad b = \$120.$$

The equation is $C(t) = 40t + 120$. Based on it we enter answers in **Table 2.19**.

e) *Can you use the equation to find out the cost of the 2-hour repair? If so, do it.*

We substitute $t = 2$ hrs into the equation and get

$$C(t) = 40t + 120, \qquad C(2) = (40)(2) + 120, \qquad C(2) = \$200.$$

f) *If a repair cost \$400, how much time did the repair take?*

We know the output value $C(t) = \$400$ and we substitute this value into the equation and solve for t:

$$400 = 40t + 120, \qquad 280 = 40t, \qquad t = 7 \text{ hrs.}$$

In section 1.5 we said that a function equation can be used as a model for a real world situation. If the equation is linear we call it a linear model.

g) *Give examples of at least two other questions that you can ask about the situation and answer using the linear model and show how to find the answers.*

What is the initial payment for a repair job? The answer is $C(0) = \$120$. Also we can ask what is the maximum cost for a one-day job. Assuming that a one-day job is at most 8 hours, the maximum cost will be $C(8) = (40)(8) + 120 = \$440$.

h) *Give an example of a question that cannot be answered using this linear model.*

How many hours it takes for a repair job of \$100? By substituting $C(t) = \$100$ into the equation and solving for t we get

$$100 = 40t + 120, \qquad -20 = 40t, \qquad t = -0.5 \text{ hrs.}$$

Since a negative time is not possible, this question cannot be answered. This makes sense as the initial charge is \$120, which is more than \$100.

i) *What are the domain and range of the model? Explain.*

Since 1.5 hours is one of the given values, we can assume that the repair job cost is paid in half-hour increments. Hence, the variables are discrete and based on the previous calculations, we have

$$\text{Real-world domain} = \{0, 0.5, 1, 1.5, ..., 8\} \text{ hrs.}$$
$$\text{Real-world range} = \{120, 140, 160, 180, ..., 440\} \text{ \$.}$$

Keys to Linear Modeling

Identify (choose) the independent and dependent variables.

Verify that the relation is linear (verbal, equation, table, graph).

Find the constant rate of change (slope) and the initial value ('y'-intercept).

Write a linear equation and give a real-world domain and a range.

Table 2.20

More Linear Modeling

For each linear situation do the tasks in the **Table 2.20**, use the equation to answer the questions, and express answers as ordered pairs or in function notation.

Example 2.4 A situation is given by the table below.

t time on elevator ride (sec)	$H(t)$ height above ground (ft.)
2	192
5	120
7	72

a) How high above ground level is the elevator after 3 seconds?

b) How long will it take to reach 10 feet above ground?

c) How high above ground level is the elevator after 12 seconds? Interpret your answer. Discuss with classmates.

Δt	t	$H(t)$	$\Delta H(t)$
+3	2	192	$\searrow$ −72
+2	5	120	$\searrow$ −48
	7	72	
	3	168	
	9.6	10	
	10	0	
	0	240	
	12	−48 ?	

Example 2.4 Extending a linear table for $H(t) = 240 - 24t$.

Solution:

The table is linear as the successive rates of change are the same

$$m = \frac{\Delta H(t)}{\Delta t} = \frac{-72}{+3} = \frac{-48}{+2} = -24 \text{ ft per sec,}$$

where DV: $H(t)$ - height of elevator (ft) and IV: t - time of ride (sec). By substituting $m = -24$ and $(7, 72)$ into the equation $H(t) = mt + b$, we get

$$72 = (-24)(7) + b, \qquad b = 240, \qquad H(t) = 240 - 24t.$$

The elevator reaches the ground floor when $H(t) = 0 = 240 - 24t$ or $-240 = -24t$ or $t = 10$ sec. If we assume that the ride ends, the domain is $[0, 10]$ sec and the range is $[0, 240]$ ft (continuous variables).

a) The height is $H(3) = 240 - 24(3) = 168$ ft.

b) Set $H(t) = 10 = 240 - 24t$ and solve for time: $-230 = -24t$ or $t \approx 9.6$ sec.

c) The height is $H(12) = 240 - 24(12) = -48$ ft which would be underground. This value is not a valid output given our previous assumption, but if the elevator goes underground, the answer could be valid. In this case we need to adjust the domain and range of our model.

Example 2.5

After a hurricane in Springfield, there was 6 feet of water covering Main Street. Each day after the hurricane the water level went down by 6 inches.

a) How long after the hurricane was there only $4\frac{1}{2}$ feet of water?

b) How deep was the flood water after 54 hours?

Solution: This situation describes a linear relation since the rate 6 inches each day is a constant rate of change, $m = -6$ in/day (decreasing), where the output is the water level L in inches and the input is the time t days after the storm. The initial value b is also given in the problem, but we must convert it to $b = 6 \times 12 = 72$ inches. So the equation is

$$L = L(t) = mt + b = -6t + 72.$$

To find the real-world domain, we must find when the water level is zero:

$$0 = -6t + 72, \qquad -72 = -6t, \qquad t = \frac{-72}{-6} = 12 \text{ days.}$$

So the domain for this model is $[0, 12]$ days and the range is $[0, 72]$ inches (continuous intervals). Recall that intervals have left end < right end.

a) We convert 4.5 feet into $4.5 \times 12 = 54$ inches, we substitute this value for L into the equation, and solve for the time t:

$$54 = -6t + 72, \qquad -18 = -6t, \qquad t = \frac{-18}{-6} = 3 \text{ days.}$$

As an ordered pair, the answer is (3 days, 54 inches).

b) We convert 54 hours into $\frac{54}{24} = 2.25$ days and evaluate the function:

$$L(2.25 \text{ days}) = -6(2.25) + 72 = 58.5 \text{ inches.}$$

Example 2.6 A situation is given by the graph below.

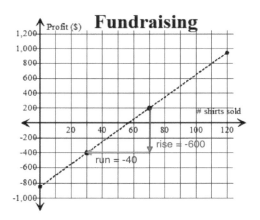

a) Find and interpret the exact 'x'-intercept.

b) Find and interpret the exact 'y'-intercept.

c) Exactly how many shirts did they sell if the profit is \$575?

d) Exactly how much profit do they earn if they sell 113 shirts?

Solution: Since the graph is a line, we have a linear model. Let s be the number of shirts sold as the independent variable and $P(s)$ the profit made in \$ as the dependent variable. By using a slope triangle, the slope is

$$m = \frac{\text{rise}}{\text{run}} = \frac{-600}{-40} = \$15 \text{ per shirt.}$$

To find the initial value we identify a lattice point on the graph, say $(70, 200)$, and substitute this point into the equation:

$$P(s) = ms + b, \qquad 200 = (15)(70) + b, \qquad b = -\$850.$$

The equation is $P(s) = 15s - 850$. From the graph, the domain is the set of values $\{0, 1, 2,...,120\}$ shirts and the range is the corresponding set of values $\{-\$850, -\$835, -\$820,...,\$950\}$ (discrete variables - dotted line).

a) To find the s-intercept, we set $P(s) = 0$ and solve the equation for s:

$$P(s) = 15s - 850, \qquad 0 = 15s - 850, \qquad s = \frac{850}{15} \approx 57 \text{ shirts.}$$

☞ A **break-even** point occurs when no money is lost or gained, **profit = \$0**.

The s-intercept is (57 shirts, \$0) and it represents the break-even point.

b) The $P(s)$-intercept is (0 shirts, $-\$850$) and shows a *loss* of \$850 if no shirt is sold. (As a side note, the purchase price was \$850/120 = \$7.08 per shirt, and the selling price was \$7.08 + \$15 = \$22.08 per shirt.)

c) We set $P(s) = 575$ in the equation and solve for s:

$$575 = 15s - 850, \qquad 1425 = 15s, \qquad s = 95 \text{ shirts.}$$

The answer is (95 shirts, \$575) or $P(95 \text{ shirts}) = \$575$.

d) We evaluate the function at $s = 113$ and get

$$P(113 \text{ shirts}) = 15(113) - 850 = \$845.$$

Example 2.7 In 2012 the average price of tickets at the box offices in the U.S. was \$7.96, and was \$8.65 in 2016. Assume that the average ticket price is a linear function of time.

a) Estimate the average ticket price in 2014.

b) What does this data predict the average ticket price would have been in the year 2017?

c) According to this model, in what year was the average price \$5.89?

d) According to this model, what would have been the average ticket price in 1965?

e) Do you think a linear model for ticket prices over time is valid? Explain/Discuss.

Solution: For this linear model the independent variable is the number of years t since 2012 and the dependent variable is the average ticket price

$A(t)$ at the box office in \$. The two data points are (0, \$7.96) and (4, \$8.65) since from 2012 to 2016 there are 4 years. The rate of change is

$$m = \frac{\text{Change in price}}{\text{Change in time}} = \frac{\$8.65 - \$7.96}{4 \text{ yrs} - 0 \text{ yrs}} = \frac{\$0.69}{4 \text{ yrs}} = \$0.1725 \text{ per year.}$$

The initial value is $b = \$7.96$ and thus the equation is

$$A(t) = 0.1725t + 7.96.$$

A real-world domain for this model could be the interval $[0, 4]$ yrs and the corresponding range could be the interval $[\$7.96, \$8.65]$. For rounding purposes, we consider both variables to be continuous. However the cutoffs for domain and range may vary.

a) Since 2014 is between 2012 and 2016, we say that we "interpolate"

$$A(2 \text{ yrs}) = 0.1725(2) + 7.96 \approx \$8.31.$$

b) The year 2017 is outside the data and we say that we "extrapolate"

$$A(5 \text{ yrs}) = 0.1725(5) + 7.96 \approx \$8.82.$$

c) To find the year we set $A(t) = 5.89$ into the equation and solve for t:

$$5.89 = 0.1725t + 7.96, \qquad -2.07 = 0.1725t, \qquad t = -12 \text{ yrs.}$$

The negative sign means 12 years *before* 2012 and so the answer is the year 2000. This is an "extrapolation" of data going back in time.

d) Since 1965 corresponds to $t = 1965 - 2012 = -47$ yrs, the average price in that year would have been

$$A(-47) = 0.1725(-47) + 7.96 \approx -\$0.15.$$

The model gives an invalid output for the year 1965.

e) Based on history, we cannot predict that ticket prices will continue to be linear over a long period of time as we may get inaccurate or invalid outputs.

Wrapping Up

In this section we have seen how to create linear models by using two data points in situations, tables, and graphs, and by writing their equations. Also we have seen how to use these equations to answer questions and how to analyze various assumptions and limitations to find valid inputs and outputs. Linear models are used in statistics to analyze real data and play an important role in economics, science, and technology. Now you learned the algebra you need to understand linear models and their many applications.

Exercises

Exercises for 2.3 Creating and Using Linear Models
(y = mx + b)

P2.10 Window washers have just finished and they begin to hoist the platform down the side of a building. The function $h(t) = 168 - 1/2t$ gives the height in feet off of the ground of the pulley-bucket (platform) t seconds after they began to hoist it down.

    a) Give one solution to this equation and explain what it means in the context of the situation.

    b) How many minutes will it take for the platform to reach the 6th floor which is 72 feet high? How many whole minutes and seconds is that?

    c) How many feet above the ground level is the platform after 4 minutes?

    d) Give an example of a question about this situation for which this linear model does not have a valid answer. Explain.

P2.11 There is a linear relationship between the Fahrenheit and Celsius degrees of measure for temperature. Water boils at 212° F and 100°C, and it freezes at 32°F and 0°C.

    • Identify (choose) the independent and dependent variables.

    • Find the constant rate of change (slope) and the initial value ('y'-intercept) for your variables.

    • Write a linear equation to represent the conversion from one temperature unit to the other.

    a) If it is 70°F outside, what is the temperature in Celsius degrees?

    b) If it is 10°C outside, what is the temperature in Fahrenheit degrees? What would you wear?

    c) Write an equation that describes the relationship between Fahrenheit and Celsius degrees.

    d) Explain the contextual meaning of the slope in your equation.

    e) Explain the contextual meaning of the 'y'-intercept in your equation.

P2.12 There is a linear relationship between the total cost of a renting a van from the Wrecks-R-Us Company and the number of miles the van is driven. Steve rented a van and paid $42.50 for driving 50 miles. Gabriella paid $53 for driving 120 miles.

    a) How much would it cost to rent a van if you drive 60 miles?

    b) How far did Nancy drive the van if it cost her a total of $72.05?

    c) Explain the contextual meaning of the slope in your equation.

    d) Explain the contextual meaning of the 'y'-intercept in your equation.

P2.13 A pool was full and the owners decided it was time to drain it for the winter. The pool drains at a constant rate, and after a half hour of draining there were still 12,150 gallons in the pool. After 9 hours of draining there were 4500 gallons in the pool.

    a) How many minutes will it take for there to be 9000 gallons in the pool? How many hours?

    b) How much water will be left in the pool after 6 hours and 15 minutes of draining?

    c) Do you have enough information to know much water the pool holds? Explain.

    d) How fast is the pool draining?

    e) How much time, in hours, will it take to empty the pool? How many minutes?

P2.14 Tomas sells cars and there is a linear relationship between the number of cars he sells each week and his weekly pay. One week Tomas sold 6 cars and he made $700 that week. Another week he sold 9 cars and made $850. This linear pay scheme is valid for the first 15 cars he sells in a week.

    a) How many cars did Tomas sell in a week that he earns $500?

    b) How much does Tomas earn if he sells 13 cars in a week?

c) Can Tomas earn $350 in a week? If so, how many cars did he sell? If not, explain why.

d) Can Tomas earn $425 in a week? If so, how many cars did he sell? If not, explain why.

e) Can Tomas earn $3400 in a week? If so, how many cars did he sell? If not, explain why.

P2.15 There is a linear relationship between the profit a club earns on a fundraiser selling school mugs and the number of mugs they sell. The club bought a total of 800 mugs to sell. If they only sell 50 mugs their profit will be −$240 (a loss). If they sell 500 mugs, their profit will be $1110.

a) If the club sells 75 mugs, what will be their profit?

b) How many mugs must the club sell to break even?

c) How much will the club earn if they sell 396 mugs?

d) Is there a limit to how much the club can earn on this fundraiser? If not explain why. If so, how many mugs must they sell and what is the maximum profit they can earn?

P2.16 A sample of charges for Willow's One-Day Repair Company is shown in the table below. To calculate charges she uses $\frac{1}{4}$-hour increments and rounds up.

t = time of repair work (hours)	$C(t)$ = total cost of repair work ($)
0	50
1	95
2	140
3	185

a) How much would a 30 minute repair cost?

b) How much would a 3 hour and 15 minute repair cost?

c) How long did a repair take if its total cost was $398.75?

P2.17 A sample of data is shown below. Assume the model is valid for 5 days.

t hrs since 5pm Sunday	$H(t)$ = height of the candle (in.)
2	6
4	6
11	6
26	6

a) What is the initial height of the candle according to the model?

b) When will the candle be 4 inches?

c) How tall will the candle be on Thursday at 6pm?

d) How tall will the candle be on Saturday at 5pm?

P2.18 A campus club bought 300 hats to sell at a fundraiser and reports the data below.

h = # of fundraising hats sold	$P(h)$ = amount of profit earned ($)
80	−980
115	−490
170	280
245	1330

a) What is the profit if they sell 123 hats?

b) How many hats did the club sell if the profit is $602?

c) How much did the club invest in the fundraiser?

d) How many hats must the club sell to break even?

e) Can the club earn $2400 in profit? If so, how many hats must they sell? If not, explain why.

P2.19 The table below shows how a certain candle burns. The candle is never blown out.

t = time a candle burns (hrs)	$P(t)$ = height of the candle (cm.)
2	21
4	18
8	12
11	7.5

a) How tall was the candle before it was lit?

b) How tall will the candle be after burning for 90 minutes?

c) Exactly how many minutes until the candle is 5 inches tall? How many seconds is that?

d) Can you tell if the candle will still be burning after 18 hours? How tall will the candle be after 18 hours?

P2.20 Below is a graph of a model (is it a model?) for the population of a town in the years since 2012.

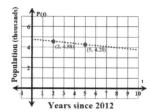

a) Interpret the meaning of the point (2, 4.58) in the context of the situation.

b) According to the model, what was the population of the town in 2015?

c) In what year does the model predict a population of about 4000?

d) Do you think this model would remain valid for a long period of time? Explain.

P2.21 Rachel is traveling and the graph is a model of her trip.

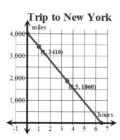

a) How far is Rachel from New York after 2 hours of travel?

b) How far is she from New York after 150 minutes of travel?

c) How much time was Rachel traveling when she was 3875 miles from New York? In hours? In minutes?

d) Where do you think Rachel was traveling from? Explain.

e) What mode of transportation do you think Rachel was taking? (car, train, plane, bus, bike, walking?) Justify your answer.

f) If Rachel began her trip at 8 am, what time did she arrive in New York?

P2.22 At a huge college graduation party a keg gets emptied at a steady pace as shown in the graph. Each person at the party will be poured one pint.

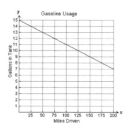

a) How many pints will have been served after 14 minutes?

b) How much time will it take to serve 67 pints?

c) How many pints will have been served after 90 seconds?

d) How many pints were in the keg when it was full?

P2.23 The gas usage for a car is shown in the graph.

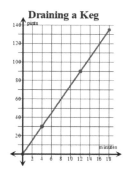

a) Interpret the slope of the line in the context of the situation.

b) Find and interpret the x-intercept and the y-intercept in the context of the situation.

c) How many miles have been driven if there are 10.2 gallons left in the tank?

d) How many miles can be driven in this car using 2 gallons of gas?

e) How many gallons will be left after driving 260 miles?

f) How far can be driven in this car on one tank of gas?

Unit 3

Exponential Models

A Review of Main Ideas

Before we begin this unit, summarize the characteristics of *linear* functions.

- Explain how you recognize a linear relationship in situations, tables, graphs, and equations.

- How do you know when the function is increasing/decreasing?

- Does the function have x- and y-intercepts? How many can it have?

Give examples and illustrate using each representation.

3.1 A New Pattern - Introduction to Exponential Functions

Launch Exploration

A scientist starts with a single bacteria cell in a petri dish (recall this dish?) After every day, each cell in the dish splits into two cells.

What should we ask first? What are the variables in this situation? How many variables are in a function? Which one is the independent variable and which one is the dependent variable? We can let

IV (Input):	d = the number of days elapsed
DV (Output):	$C(d)$ = the total amount of cells in dish after d days

What are the constants in this situation? What is the initial value? What is the output value in the next day? Assume that this 'Malthusian' model works for one month. Without using a calculator predict the number of cells after one month. Estimate for fun, be brave! Now get together with a group of classmates and do the tasks in **Table 3.1** on the margin except for the equation.

Keys to Modeling

Identify the dependent vs. independent variables including units.

Make a table of values.

Examine the table for a pattern.

Linear? EXPLAIN.

Write an equation (formula) to represent the function.

Table 3.1

Exponential Function Tables

The variables d, $C(d)$ in the **Launch Exploration** are discrete as they are counted:

IV (Input): d = the number of days elapsed (unit: days)

DV (Output): $C(d)$ = the total number of cells in dish after d days (unit: cells)

Table 3.2 shows that the function is definitely not linear since the first differences are not constant. However, there is a *constant multiplier* (*factor*) of 2, as we can see in **Table 3.3**. That means to find the number of cells on any given day, we multiply the number of cells from the previous day by a constant factor of 2.

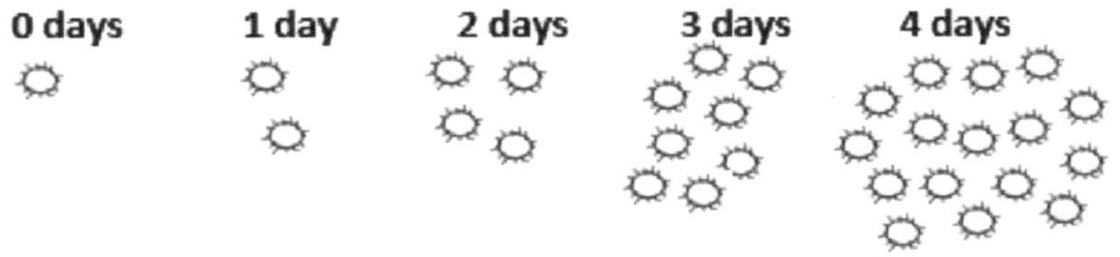

Diff	d	$C(d)$	1ˢᵗ Diff
+1	0	1	+1
+1	1	2	+2
+1	2	4	+4
+1	3	8	+8
	4	16	

Table 3.2 Linear? - No

Diff	d	$C(d)$	Mult
+1	0	1	×2
+1	1	2	×2
+1	2	4	×2
+1	3	8	×2
	4	16	

Table 3.3 Constant multiplier/factor (Exponential? - Yes)

Figure 3.1 Cell doubling.

> **Definition 3.1 *Exponential Function Tables***
>
> **Exponential function tables** are tables of values characterized by **constant multipliers** of the output *when the input increases by a constant.*

The constant multiplier takes us from output to output going forward in the exponential table

(3.1)   **current output** × multiplier = **next output**

and it is obtained by taking the ratios of outputs going backwards as in **Table 3.4**

(3.2)   multiplier = **current output/previous output**.

d	$C(d)$	Mult = Ratio
0	1	×2 = $\frac{2}{1}$
1	2	×2 = $\frac{4}{2}$
2	4	×2 = $\frac{8}{4}$
3	8	×2 = $\frac{16}{8}$
4	16	

Table 3.4 Finding multipliers

> **Definition 3.2 *Multipliers or Successive Ratios***
>
> In a table of values of a function $y = f(x)$, the **multipliers** or the **successive ratios** of the **output y** are defined by
>
> | x | y | Multiplier = Ratio |
> |---|---|---|
> | x_0 | y_0 | × y_1/y_0 |
> | x_1 | y_1 | × y_2/y_1 |
> | x_2 | y_2 | × y_3/y_2 |
> | x_3 | y_3 | |

Example 3.1

For the **Table 3.5**, answer the following questions:

- Does the table represent an exponential function? Explain why or why not.
- IF it DOES represent an exponential function, identify the constant multiplier.

d	$N(d)$
0	24,000
1	30,000
2	37,500
3	46,875

Table 3.5

Solution: The successive output ratios are constant (all the same) and equal to 1.25 as the input increases by one unit. So the table represents an exponential function with the constant multiplier = 1.25.

$$\frac{N(1)}{N(0)} = \frac{30,000}{24,000} = 1.25, \qquad \frac{N(2)}{N(1)} = \frac{37,500}{30,000} = 1.25, \qquad \frac{N(3)}{N(2)} = \frac{46,875}{37,500} = 1.25.$$

(In this table the output values are increasing.)

Diff	d	$N(d)$	Mult
+1	0	24,000	×1.25
+1	1	30,000	×1.25
+1	2	37,500	×1.25
	3	46,875	

Example 3.1 Exponential? - Yes

Example 3.2

For the **Table 3.6**, answer the following questions:

- Does the table represent an exponential function? Explain why or why not.
- IF it DOES represent an exponential function, identify the constant multiplier.

t	$P(t)$
0	28,000
1	21,000
2	14,000
3	7,000

Table 3.6

Solution: The successive output ratios are:

$$\frac{P(1)}{P(0)} = \frac{21,000}{28,000} = 0.75, \qquad \frac{P(2)}{P(1)} = \frac{14,000}{21,000} = \frac{2}{3} \approx 0.67.$$

No, this table does NOT represent an exponential function because the consecutive output values do not decrease by a common multiplier although the inputs increase by one unit - there is no common ratio between consecutive outputs.

Diff	t	$P(t)$	Mult
+1	0	28,000	×0.75
+1	1	21,000	×0.67
+1	2	14,000	×0.50
	3	7,000	

Example 3.2 Exponential? - No

Exponential Function Equations

Going back to the **Launch Exploration**, let $d = 0, 1, 2, 3, \ldots$ days and observe the pattern displayed by the number of cells below:

$$C = C(0) = 1 \qquad\qquad = 1 \cdot 2^0 \qquad\qquad \text{(Initial Value)}$$

$$C = C(1) = (1) \cdot 2 \qquad\qquad = 1 \cdot 2^1 \qquad\qquad (\times 2)$$

$$C = C(2) = (1 \cdot 2) \cdot 2 \qquad\qquad = 1 \cdot 2^2 \qquad\qquad (\times 2)$$

$$C = C(3) = (1 \cdot 2 \cdot 2) \cdot 2 \qquad\qquad = 1 \cdot 2^3 \qquad\qquad (\times 2)$$

$$\ldots$$

$$C = C(d) = (1 \cdot 2 \cdot 2 \cdot \ldots \cdot 2) \cdot 2 \qquad\qquad = 1 \cdot 2^d. \qquad\qquad \text{(Exponential Function)}$$

We conclude that the equation of the cell growth function is

(**3.3**) $C(d) = 1 \cdot 2^d$ cells, $d = 0, 1, 2, 3, ..., 30$ days.

Now that we have the equation, we can find the number of cells after 30 days:

$$C(30) = 1 \cdot 2^{30} = 1,073,741,824 \qquad \text{– over 1 billion cells!}$$

What does "exponential growth" mean to you? How close was your estimate to the true value? Most people make estimates that are nowhere near the actual value! To find the true value we clearly need a formula and a calculator.

Exponential function *equations* (*formulas*) can always be written in the form:

$$y = f(x) = a \cdot b^x, \qquad\qquad \text{where } b > 0.$$

Exponential functions are given by *exponentiating* a constant b called the *base* of the exponential with the input variable x as the *exponent*. The constant a is the *initial value* of the function. It represents the value of the output y when the input x has a value of 0:

$$y = f(0) = a \cdot b^0 = a \cdot 1 = a.$$

If we evaluate the function for $x = 1, 2, 3, ...$ we can see that b is the *constant multiplier* or *factor* of the function *when the input increases by one unit*:

$$y = f(1) = a \cdot b^1 = a \cdot b = ab,$$
$$y = f(2) = a \cdot b^2 = a \cdot b \cdot b = abb,$$
$$y = f(3) = a \cdot b^3 = a \cdot b \cdot b \cdot b = abbb.$$

☞ **Why is it that b cannot be negative or zero?**

If $b = -1$ and $x = 1/2$, then

$y = b^x = (-1)^{1/2} = \sqrt{-1}$

is not a real number.

If $b = 0$ and $x = 0$, then $b^x = 0^0$ is not even defined!

☞ **What happens if $a = 0$ or $b = 1$?**

If $a = 0$, the function is

$y = a \cdot b^x = 0 \cdot b^x = 0$.

If $b = 1$, the function is

$y = a \cdot b^x = a \cdot 1^x = a \cdot 1 = a$.

In both cases the function is a constant function.

Definition 3.3 *Exponential Function Equation (Formula)*

An **exponential function** is defined by an equation (formula) of the form

$$y = f(x) = a \cdot b^x, \qquad a \neq 0, \qquad 1 \neq b > 0.$$

where x is the independent variable (input), y is the dependent variable (output), and a, b are constants. Moreover, a represents the **initial value**, and b the **constant multiplier** or **factor** of the function.

Examples of Exponential Functions

The following equations (formulas) define exponential functions:

☞ Recall that $b^{m \cdot n} = (b^m)^n$

$2^{-x} = 2^{(-1) \cdot x} = (2^{-1})^x$

$3^{2s} = 3^{2 \cdot s} = (3^2)^s$.

$y = 7^x$	$a = 1,$	$b = 7$
$f(x) = 5 \cdot 2^{-x}$	$a = 5,$	$b = 2^{-1} = \frac{1}{2}$
$h(s) = -40 \cdot 3^{2s}$	$a = -40,$	$b = 3^2 = 9$
$y = 4.9 \cdot (1.05)^t$	$a = 4.9,$	$b = 1.05$

Non-Examples of Exponential Functions

The following equations (formulas) do not define exponential functions:

$y = 7x^2$ — Not exponential because the variable x is in the base, not in the exponent. This is a quadratic.

$V(r) = 3r - 8$ — Not exponential for the same reason as above. (What kind of function is this?)

$y = 20(1.2)^t + 5$ — Well, this is a *vertical shift* of an exponential function, $20(1.2)^t$, but it is not an exponential.

$P(t) = 32 \cdot (-1.05)^t$ — Not exponential since the base -1.05 is not positive.

More About Writing Equations from Exponential Function Tables

When a table of values is an exponential function table, a natural question is "What is the equation (formula) of the function?" According to **Definition 3.3**, this equation should be of the form

$$y = f(x) = a \cdot b^x$$

where $a = f(0)$ is the initial value and b is the constant multiplier when the input increases by one unit. In **Example 3.1** we can read from the **Table 3.5** the initial value $N(0) = 24,000$ and we already found the constant multiplier 1.25 when the input increases by one unit. So $N(d) = a \cdot b^d$ where $a = 24,000$ and $b = 1.25$. In other words, the equation is

$$N(d) = 24,000 \cdot (1.25)^d.$$

If the table does not have a constant multiplier, the function is not exponential and cannot be given by an exponential formula. Always check first for the common multiplier.

Diff	d	N(d)	Mult
+1	0	24,000	×1.25
+1	1	30,000	×1.25
+1	2	37,500	×1.25
	3	46,875	

Example 3.1

☞ The independent variable is d and the dependent variable is $N(d)$.

d	N(d)
0	45,000
1	30,000
2	20,000
3	13,333

(a)

x	y
3	4
4	1
5	0.25
6	0.0625

(b)

t	P(t)
7	24,000
8	30,000
9	37,500
10	46,875

(c)

Table 3.7 Exponential?

Verify the table is exponential.

Find the constant multiplier of the function.

Find the initial value of the function.

Write the equation (formula) of the function.

Example 3.3

Example 3.3

For each table of values in **Table 3.7**, do the tasks in the margin.

Diff	d	$N(d)$	Mult
+1	0	45,000	×(2/3)
+1	1	30,000	×(2/3)
+1	2	20,000	×(2/3)
	3	13,333	

Example 3.3 a)
$N(d) = 45,000 \cdot (0.67)^d$.

Diff	x	y	Mult
+1	3	4	×0.25
+1	4	1	×0.25
+1	5	0.25	×0.25
	6	0.0625	

Example 3.3 b) $y = 256 \cdot (0.25)^x$.

Diff	x	y	Div
−1	3	4	÷0.25
−1	2	16	÷0.25
−1	1	64	÷0.25
	0	256	

Example 3.3 b) Reverse Table.

☞ Division is multiplication by reciprocal:

$$\frac{4}{0.25} \div (0.25) = \frac{4}{0.25} \cdot \frac{1}{0.25}$$
$$= \frac{4}{(0.25)^2}.$$

Diff	t	$P(t)$	Mult
+1	7	24,000	×1.25
+1	8	30,000	×1.25
+1	9	37,500	×1.25
	10	46,875	

Example 3.3 c)
$P(t) = 5,033.16 \cdot (1.25)^t$.

Solution: Notice that in each table the *inputs* increase by *one unit* and we have to search first for a common multiplier for the *outputs*.

a) The successive output ratios are constant (all the same) and equal to 2/3. So the table represents an exponential function with the constant multiplier:

$$b = \frac{30,000}{45,000} = \frac{2}{3}.$$

We read the initial value directly from the table: $a = N(0) = 45,000$. As the independent variable is d, the exponential function equation (formula) is

$$N(d) = a \cdot b^d = 45,000 \cdot \left(\frac{2}{3}\right)^d.$$

b) The successive output ratios are constant and equal to 0.25. So the function is exponential with the constant multiplier:

$$b = \frac{1}{4} = 0.25.$$

The initial value $y = a$ of this function is when $x = 0$. As that value is not listed in the table, it is necessary to work backwards, which means that it is necessary to DIVIDE by the multiplier instead of multiply by it. More specifically, we go backwards in the table from the point $(3,4)$ to the point $(0,a)$ by successively dividing the output value 4 by the multiplier 0.25 three times:

$$(3,4) \to (2,16) \to (1,64) \to (0,256); \qquad a = 256.$$

Another way to view this set of data is to write a reverse table in the margin to eventually arrive at the initial output when the initial input $x = 0$. To see the pattern, we can write the sequence above using algebra as follows:

$$(3,4) \to \left(2, \frac{4}{0.25}\right) \to \left(1, \frac{4}{(0.25)^2}\right) \to \left(0, \frac{4}{(0.25)^3}\right); \qquad a = \frac{4}{(0.25)^3} = 256.$$

As the independent variable is x, the exponential function equation (formula) is

$$y = a \cdot b^x = 256 \cdot \left(\frac{1}{4}\right)^x = 256 \cdot (0.25)^x.$$

c) The successive output ratios are constant and all equal to 1.25. So the function is exponential with the constant multiplier:

$$b = \frac{30,000}{24,000} = \frac{5}{4} = 1.25.$$

To find the initial value $P(0) = a$, we go backwards in the table from the point $(7, 24000)$ to the point $(0, a)$ by dividing the output value 24000 by the multiplier 1.25 seven times:

$$a = \frac{24,000}{(1.25)^7} = 5,033.16; \qquad P(t) = a \cdot b^t = 5,033.16 \cdot (1.25)^t.$$

An alternative method to find the initial value

To write equations for the exponential tables in **Example 3.3** we first found the constant multipliers by taking ratios (current output)/(previous output). Next we found the initial value of each function by using the constant multiplier as a divisor to go backwards in the table until we got the output for input 0. But if we know the constant multiplier and any point on the graph (or in the table) we can find the initial value by another method - substitution.

☞ If $(0, a)$ is already in the table, the initial value is easy to find by reading the table.

Example 3.3 b). The constant multiplier is $b = 0.25$ and thus, the exponential function equation is of the form $y = a \cdot b^x = a \cdot (0.25)^x$. Each entry in the table of values is an ordered pair solution (x, y) of this equation, so we can substitute the values of any of these points in the equation. For example $(x, y) = (3, 4)$ gives

$$4 = a \cdot (0.25)^3, \qquad a = \frac{4}{(0.25)^3} = 256, \qquad y = 256 \cdot (0.25)^x. \qquad \text{(Substitution)}$$

If we choose $(x, y) = (4, 1)$ we get $1 = a \cdot (0.25)^4$ and again $a = 1/(0.25)^4 = 256$.

Example 3.3 c). Using the same method, we can substitute the values $(t, P(t)) = (7, 24000)$ into the equation $P(t) = a \cdot (1.25)^t$ to solve for a:

$$P(7) = 24000 = a \cdot (1.25)^7, \qquad a = \frac{24000}{(1.25)^7} = 5033.16, \qquad P(t) = 5033.16 \cdot (1.25)^t. \qquad \text{(Substitution)}$$

Writing Equations from Two (Consecutive) Data Points

If we already know that the function is exponential, then we only need *two* consecutive data points to find its multiplier/factor b and use either a reversed table or the substitution method to find its initial value a.

☞ Here by 'consecutive' points we mean that their inputs increase by one unit.

Example 3.4

Find the equation of an exponential function whose graph/table contains the points $(4, 117)$ and $(5, 218)$.

Solution: The two points are consecutive as the inputs from 4 to 5 increase by one. The multiplier/ factor is thus given by $b = 218/117 \approx 1.86$. To find the initial value a we write the equation $y = a \cdot b^x = a \cdot (1.86)^x$ and substitute the data point $(x, y) = (4, 117)$ into this equation:

Diff	x	y	Mult
+1	4	117	$\times 1.86$
	5	218	

Example 3.4

$$117 = a \cdot (1.86)^4, \qquad a = \frac{117}{(1.86)^4} \approx 33.82, \qquad y = 33.82 \cdot (1.86)^x. \qquad \text{(Substitution)}$$

Example 3.5

A 20-pound ice block melts exponentially. Find an equation for the weight $W(t)$ of the block in pounds after t hours of melting if it weighs 15 pounds after one hour.

t hrs	$W(t)$ lbs	Mult
0	20	$\succ \times 0.75$
1	15	

Example 3.5

t yrs	$V(t)$\$	Mult
1	200,000	$\succ \times 1.15$
2	230,000	

Example 3.6

(Substitution)

Things to remember about the Exponential Equation $y = a \cdot b^x$

The input variable (x) is in the exponent.

The base (b) $\neq 0, 1$ and is not negative.

The initial value (a) is not zero.

Solution: In this situation, the independent variable is time t in hours and the dependent variable is the weight $W(t)$ of the block in pounds after t hours. We are given two consecutive data points $(0, 20)$ and $(1, 15)$ of an exponential function. The initial value, the multiplier/factor, and the equation are

$$a = W(0) = 20, \qquad b = \frac{15}{20} = \frac{3}{4}, \qquad W(t) = a \cdot b^t = 20 \cdot \left(\frac{3}{4}\right)^t.$$

Example 3.6

The median house value in Springfield grew exponentially during a real estate bubble. After the first year it was \$200,000 and increased to \$230,000 the next year. Write an equation for the median value $V(t)$ of a house in dollars after t years.

Solution: We know that $V(t)$ is an exponential function of t with two consecutive data points $(1, 200000)$ and $(2, 230000)$. The multiplier/factor is $b = 230000/200000 = 1.15$. To find the initial value a we write the equation

$$V(t) = a \cdot b^t = a \cdot (1.15)^t$$

and substitute the data point $(t, V(t)) = (1, 200000)$ into this equation:

$$200,000 = a \cdot (1.15)^1, \quad a = \frac{200,000}{(1.15)^1} = 173,913.04, \quad V(t) = 173,913.04(1.15)^t.$$

Wrapping Up

By modeling a cell-growth situation, we discovered a new type of function - the exponential. Exponential function tables are characterized by two properties: 1) inputs increase by a constant and 2) outputs change by a constant multiplier or factor. This multiplier is the ratio of any output to its previous output going backwards in the table.

Exponential function equations (formulas) can always be written in the form $y = f(x) = a \cdot b^x$. To write this equation we need the two constants: the base b and the initial value $a = f(0)$. To find them we only need two consecutive data points, i.e., two points with inputs increasing by one. The constant multiplier, b, is the ratio of the later output to the previous output. The initial value is obtained by substituting one of the two points (x, y) and the multiplier b into the equation $y = a \cdot b^x$ and solving for a.

The variable x, the base b, and the initial value a must satisfy the conditions written in the margin in order for the formula to define a (non-constant) exponential function. In real-world situations the initial value a is usually positive and in the next section we will classify exponential functions as growth or decay and examine the characteristics of their graphs.

Exercises

Exercises for 3.1 A New Pattern - Introduction to Exponential Functions

P3.1   For each (exponential) situation:

- Identify and label the two related variables including their units. Which is independent? Dependent?

- Identify the initial value and the constant multiplier.

- Write an equation for the function that relates the two variables.

a) A lab started with 50 grams of a radioactive substance. Each week only half of the previous week's radioactive amount remained.

b) Marcus' parents put $6,000 into an account when he was born and then didn't touch it. At the end of each decade the account contained 130% of the amount from a decade earlier.

c) A small software company was worth $100,000 when it began selling video games. After every month its value was 9/7 times the previous month's value.

d) A website got 8,000,000 hits in its first month. Every month after that the site got only 5/6 of the number of hits from the month before.

e) Every month after a factory was abandoned the population of cockroaches inside was 1.4 times the population from the month before. The population of cockroaches was about 40,000 when the factory was first abandoned.

P3.2   Read about Thomas Malthus (1766 - 1834). Explain why the model $C(d) = 2^d$ for cell growth is called 'Malthusian'. Explain why it cannot work forever in the real-world.

P3.3   For each of the following tables do the following tasks:

- Does the table represent an exponential function? Explain why or why not.

- IF it DOES represent an exponential function, identify: 1) the initial value; 2) the constant multiplier/factor.

- Write an equation for the function. Use it to fill in the missing values in the table.

a)

d days	$N(d)$ cells
0	24,000
1	30,000
2	37,500
3	46,875
10	
15	

b)

d days	$N(d)$ cells
0	45,000
1	30,000
2	20,000
3	13,333
10	
15	

c)

t years	$P(t)$ people
0	220,000
1	231,000
2	242,550
3	254,677.5
10	
15	

d)

t years	$P(t)$ people
0	28,000
1	21,000
2	14,000
3	7,000
10	
15	

e)

d days	$N(d)$ cells
0	24,000
1	18,000
2	13,500
3	10,125
10	
15	

P3.4   Identify which functions are exponential and which are not and explain why. For those that *are* exponential,

- Identity the initial value and the constant multiplier/factor.
- Start a table of values without a calculator.

a) $y = 64 \cdot (1.25)^x$

b) $A(t) = 27{,}000 \cdot \left(\frac{1}{3}\right)^t$

c) $y = 6(1)^x$

d) $N(x) = 1{,}000{,}000 \cdot (0.6)^x$

e) $A(t) = 100{,}000 \cdot \left(\frac{5}{4}\right)^{-t}$

f) $y = 10 \cdot (0.35x)^2$

g) $N(x) = 4000 \cdot (1.045)^x$

h) $f(t) = 1600 \cdot (4)^{-t}$

i) $y = 20 \cdot (-1.4)^x$

j) $y = 0.4 \cdot (2)^{3t}$ (optional)

k) $y = 0.5 \cdot (4)^{t/2}$ (optional)

P3.5 How many points do you need in an exponential function table to write its equation?

P3.6 Write an equation for the exponential function shown in the table.

a)

d days	$N(d)$ cells
2	81000
3	54000
4	36000

b)

d days	$N(d)$ cells
3	24000
4	36000
5	54000

c)

x	y
4	200
5	600
6	1800

d)

x	$F(x)$
8	8,192
9	32,768

P3.7 For each of the following tables, determine whether the pattern in the table represents a linear, an exponential, or neither relationship.

Be prepared to justify your answer by showing HOW you recognize a linear or an exponential pattern in a table.

a)

x	y
0	400
1	200
2	100
3	50

b)

x	y
0	8100
1	2700
2	2000
3	1000

c)

x	y
0	0.5
1	2
2	8
3	32
4	128

P3.8 Since the Smith family moved to Littleton, the population has been growing exponentially. In 2 years it was 14,520 and after 3 years it was 15,972. Write an equation for $P(t)$, the population t years after the Smiths moved in.

P3.9 A radioactive sample was delivered to a lab. Three days later the sample was 48,514,950 milligrams and the day after that it was 48,029,800.5 milligrams. Write an exponential function equation $A(d)$ representing the amount in milligrams of the sample left after d days.

P3.10 Marcus invested some money into an account to earn annual interest and grow exponentially. After six years the amount grew to $18,018.60 and after seven years to $18,378.97. Write an equation for $A(t)$, the amount of money in dollars t years after Marcus invested.

P3.11 A car dealer told Shanya that the value of her new car will decrease exponentially after purchase. After one year it will be $22,080 and after two years it will be $20,313.60. Write an equation for the value, $V(t)$, of Shanya's car t years after purchase.

P3.12 The distance from the Earth to the Moon is about 384,000 km and the paper thickness is about 0.1 mm. Show that by successively folding a sheet of magic paper once per day you can reach the Moon in 42 days. (optional)

3.2   Features of Exponential Function Graphs

Launch Exploration

A scientist starts with a single bacteria cell in a petri dish. After every day, each cell in the dish splits into two cells. The variables are

> IV (Input): d = the number of days elapsed
> DV (Output): $C(d)$ = the number of cells in dish after d days

and the equation (formula) representing this function is $C(d) = 1 \cdot 2^d$. Now get together with a group of classmates and do the tasks in **Table 3.8** in the margin. Assume that this Malthusian model works for one month.

> **Keys to Modeling**
>
> Make a table of values.
>
> Identify the real-world domain and range.
>
> Create a graph for the function.
>
> Determine whether the function is increasing or decreasing.

Table 3.8

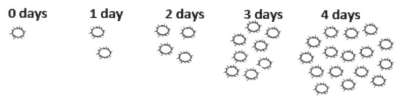

Figure 3.2 Cell doubling.

An Exponential Growth Example

We have previously established that the cell growth **Equation (3.3)** is given by $C(d) = 1 \cdot 2^d$, which is an exponential function with initial value $a = 1$ and the constant multiplier/factor $b = 2$. Based on the information given in the **Launch Exploration**, this "Malthusian model" for the cell growth works for one month. So the real-world domain of the function is the set of integers 0,1, 2, 3, ..., 30 (days), which we can also write as a list (with an ellipsis!)

(3.4)   **Domain = {0 days, 1 day, 2 days, ..., 30 days}.**

☞ Both variables d and $C(d)$ are discrete.

To find the range, we need to know the least number of cells and the greatest number of cells over the 30 days period. On day zero we have 1 cell and we double the cells each day. So the greatest number of cells occurs on day 30 and will be

> $C(30) = 1 \cdot 2^{30} = 1,073,741,824 \approx 1.07 \times 10^9$ – scientific notation!

So the range is the set of powers $2^0, 2^1, 2^2, 2^3, ..., 2^{30}$, which we can write as a list:

(3.5)   **Range = {1 cell, 2 cells, 4 cells, 8 cells, ..., 1.07×10^9 cells}.**

☞ To evaluate 1.07×10^9 we move the decimal point nine places to the right. For 1.07×10^{-9} we move it to the left nine places.

This notation is used to estimate very large or very small numbers, some of which would be impossible to write as decimals.

We conclude that the graph is the set of points represented by ordered pairs:

> Graph = {(0,1), (1,2), (2,4), (3,8), ..., $(30, 1.07 \times 10^9)$}.

However, to see the complete mathematical graph, we include some negative values for the input such as $d = -1, -2, -3$, which are not in the real-world domain. These values are in the *mathematical* domain since we do get mathematically valid outputs (See **Table 3.9**):

☞ Recall that $2^{-n} = \left(\frac{1}{2}\right)^n = \frac{1}{2^n}$.

$$C(-1) = 1 \cdot 2^{-1} = \frac{1}{2}, \qquad C(-2) = 1 \cdot 2^{-2} = \frac{1}{4}, \qquad C(-3) = 1 \cdot 2^{-3} = \frac{1}{8}.$$

For a better visual, we show a partial graph in **Fig 3.3 (a)** and the complete graph in **Fig 3.3 (b)** for our cell growth situation. In both cases, the graph consists only of the (larger) black dots since the variables are *discrete*. Here the dotted line indicates what the graph would look like if the variables were *continuous*. Deciding whether or not to connect the dots with a solid line depends on whether or not the variables are continuous. If it were continuous, it wouldn't be dotted.

$C(d) = 1 \cdot 2^d$		
d days	$C(d)$ cells	$(d, C(d))$
−3	$1/8 \approx 0.13$	$(-3, 0.13)$
−2	$1/4 = 0.25$	$(-2, 0.25)$
−1	$1/2 = 0.5$	$(-1, 0.5)$
0	1	$(0, 1)$
1	2	$(1, 2)$
2	4	$(2, 4)$
3	8	$(3, 8)$
4	16	$(4, 16)$
...	...	...
30	1.07×10^9	

Table 3.9

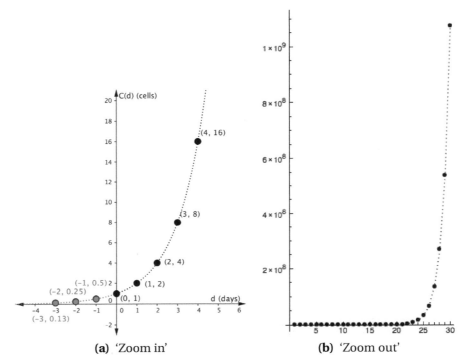

(a) 'Zoom in'   **(b)** 'Zoom out'

Figure 3.3 The red dots are not part of the real-world graph (model.)

☞ The graph of an increasing function goes up as you trace it from left to right.

Notice that drawing a complete exponential graph requires a scale for the vertical axis to handle very large numbers. That scale will strongly distort the diagram. To see this distortion, compare the 'zoom in' version **Fig 3.3 (a)**, where the y-**intercept** $(0, 1)$ is visible, and the 'zoom out' version **Fig 3.3 (b)**, where the y-intercept is not visible. This shows that although exponential functions grow slowly at the left end, they eventually grow very large very quickly at the right end. In particular, the exponential function $C(d) = 1 \cdot 2^d$ is **increasing** and the multiplier $b = 2$, which is also **> 1**, tells how many times an output value of the function is **greater** than a previous output value if the input *increases by one unit*.

The Asymptote and Intercepts

In **Fig 3.3** notice also that the graph is getting closer and closer to the x-axis at the left end, but **never crosses** or **touches** the x-axis. This is due to the fact that the values of the function $C(d) = 1 \cdot 2^d$ are always **positive**, but get very small very quickly if the values of the exponent are negative.

> **Fact 3.4** *The Asymptote*
>
> Any exponential graph of the form $y = f(x) = a \cdot b^x$ gets closer and closer to the x-axis without touching - **no x-intercepts**. We say that the graph is **asymptotic** to the x-axis or that the x-axis is an **asymptote** of the graph.

All exponential graphs also have a y-intercept described as follows:

> **Fact 3.5** *The y-Intercept*
>
> The y-**intercept** of any exponential graph of the form $y = f(x) = a \cdot b^x$ is the **point** $(0, a)$ on the graph having **zero** as its x-coordinate, $x = 0$, and the **initial value** of the function as its y-coordinate, $y = f(0) = a$.

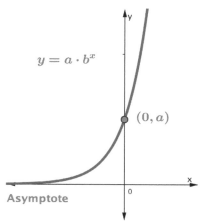

Figure 3.4 Exponential Graph

An Exponential Decay Example

A strain of bacteria starts with 200,000,000 cells. Every day 1/2 of the cells die.

IV (Input): d = the number of days elapsed (unit: days)
DV (Output): $N(d)$ = the number of cells alive after d days (unit: cells)

Table 3.10 of values for this situation is an exponential function table with the **constant multiplier**

d	$N(d)$	Mult
0	200,000,000	×0.5
1	100,000,000	×0.5
2	50,000,000	×0.5
3	25,000,000	×0.5
...	...	×0.5
?	1	

Table 3.10

$$b = 1/2 = 0.5, \quad \text{where} \quad 0.5 < 1.$$

The **initial value** of the function is given by the entry in the table with $d = 0$

$$a = N(0) = 200,000,000 = 2 \times 10^8. \qquad \text{(Scientific notation)}$$

The **equation** representing the function is

(3.6) $$N(d) = a \cdot b^d = 200,000,000 \cdot (0.5)^d. \qquad \text{(Substitution)}$$

The **real-world domain** starts with day zero and ends with the first day when the last cell starts to die. That day is estimated by the solution d to the equation

$$N(d) = 200,000,000 \cdot (0.5)^d = 1. \qquad \text{(Equation)}$$

To solve this equation by arithmetic means, we extend **Table 3.10** forward using the formula and a calculator. We find two estimates by trial and error:

$$N(27) = 200,000,000 \cdot (0.5)^{27} \approx 1.49 > 1,$$

(Evaluate)

$$N(28) = 200,000,000 \cdot (0.5)^{28} \approx 0.75 < 1.$$

So the day $d = 28$ is the first day when the last cell starts to die. After 28 days this model is no longer valid for this situation. (Why?)

(3.7)   **Domain = {0 days, 1 day, 2 days, ..., 28 days}.**

The corresponding **real-world range** is given from *least to greatest* by the list:

(3.8)   **Range = {1 cell, 2 cells, 4 cells, ..., 200000000 cells}.**

The **graph** consists of 28 points as the variables d, $N(d)$ are discrete:

$$\text{Graph} = \{(0, 2 \times 10^8), (1, 10^8), (2, 5 \times 10^7), ..., (28,1)\}.$$

$N(d) = 200,000,000 \cdot (0.5)^d$	
d	$N(d)$
0	$200,000,000 = 2 \cdot 10^8$
1	$100,000,000 = 1 \cdot 10^8$
2	$50,000,000 = \frac{1}{2} \cdot 10^8$
3	$25,000,000 = \frac{1}{4} \cdot 10^8$
4	$12,500,000 = \frac{1}{8} \cdot 10^8$
...	...
28	1

Table 3.11

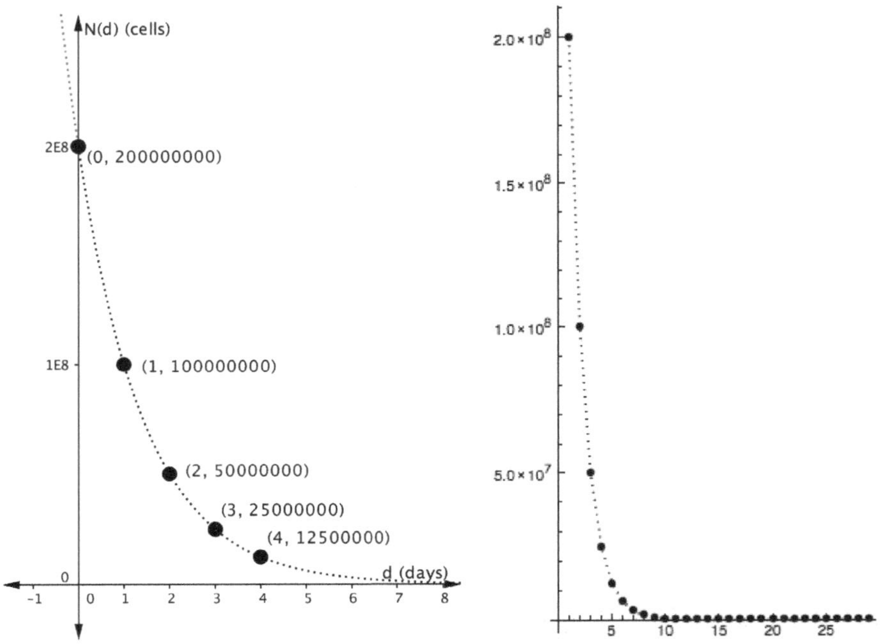

(a) 'Zoom in'   **(b)** 'Zoom out'

Figure 3.5 The red dots are not part of the real-world graph (model.)

☞ The graph of a decreasing function goes down as we trace it from left to right.

The function is **decreasing** and gets close to zero very quickly at the right end as each day the number of cells alive is 1/2 the number of cells from the previous day. The multiplier $b = 0.5$ (< 1) tells how many times an output value is **smaller** than the previous output value as the input *increases by one unit*. The y-**intercept** of the graph is $(0, a) = (0, 2 \times 10^8)$ as 2×10^8 is the initial number of cells in scientific notation. Moreover, the graph gets closer and closer to the x-axis at the right end, but **never touches** the x-axis, meaning that the x-axis is an **asymptote** of the graph. So *theoretically* there never would be zero cells.

Exponential Growth/Decay Functions

When we multiply a positive number by a factor b, the product is larger than the original number if $b > 1$ and smaller if $0 < b < 1$. Exponential functions with positive initial value remain positive and can be classified as follows:

☞ If we multiply 5 by 3 > 1, the product is 5 × 3 = 15 > 5.

If we multiply 5 by 0.3 < 1, we get 5 × 0.3 = 1.5 < 5.

Fact 3.6 *The Exponential Growth/Decay Functions*

Any exponential function $y = f(x) = a \cdot b^x$ with $a > 0$ is either

I) **Increasing for $b > 1$**, in which case we say that it is an **exponential growth** function with the **growth factor b**.

OR

II) **Decreasing for $0 < b < 1$**, in which case we say that it is an **exponential decay** function with the **decay factor b**.

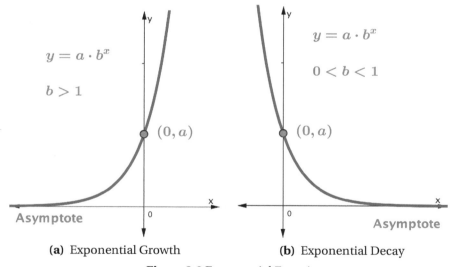

(a) Exponential Growth   **(b)** Exponential Decay

Figure 3.6 Exponential Functions

Keys to Analyzing Exponentials

Decide if the function represents growth or decay and explain.

Identify the growth/decay factor and the initial value.

Determine the real-world domain and range.

Sketch a graph, label the y-intercept and the asymptote.

Table 3.12

Exponential growth graphs increase while decay graphs decrease, but notice that they have the same general shape in reverse directions - their shapes are just reflections of each other across the vertical axis.

Analyzing Exponential Functions in Context

For each **Example 3.7**, **3.8**, **3.9** do the tasks in **Table 3.12** in the margin. The asymptotes may not be relevant for all real-world graphs.

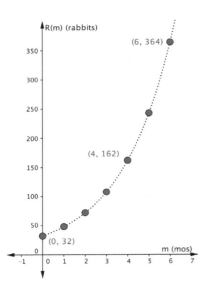

Figure 3.7 Rabbits Population
Growth

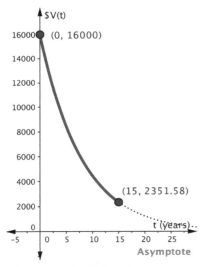

Figure 3.8 Car Value Depreciation

☞ The left end L in the interval notation $[L, R]$ must be smaller than the right end R.

Example 3.7

The function $R(m) = 32 \cdot (1.5)^m$ models a rabbit population over a six-month period where $R(m)$ is the number of rabbits after m months.

Solution: The base of the exponential is $b = 1.5 > 1$ and thus, we have an exponential growth function with the growth factor of 1.5. To find the initial value, we take the input $m = 0$ and evaluate the output

$$R(0) = 32 \cdot (1.5)^0 = 32 \cdot 1 = 32 \text{ rabbits.}$$

So the initial value is 32 and represents the number of rabbits in the initial population. The model is given over a six-month period of time. Since the number of rabbits is a *discrete* variable, it is convenient to regard the time variable m also as a discrete variable. So the real-world domain is the list

$$\text{Domain} = \{0 \text{ mos., } 1 \text{ mo., } 2 \text{ mos., } \dots , 6 \text{ mos.}\}.$$

The corresponding range is the list of output values of the function $R(m)$ for $m = 0, 1, 2, .., 6$ (rounded to the integer part):

$$\text{Range} = \{32 \text{ rabbits, } 48 \text{ rabbits, } 72 \text{ rabbits, } \dots , 364 \text{ rabbits}\}.$$

The graph consists of 6 points and a sketch is shown in the margin, where the y-intercept is $(0, 32)$ and the dotted line indicates only the trend.

Example 3.8

The value of a car over a 15-year period is given by the formula $V(t) = 16,000 \cdot (0.88)^t$ in dollars where t is the time in years since its purchase.

Solution: The base of the exponential is $b = 0.88 < 1$ and thus, we have an exponential decay function with the decay factor of 0.88. To find the initial value, we take the input $t = 0$ and evaluate the output

$$V(0) = 16,000 \cdot (0.88)^0 = 16,000 \cdot 1 = \$16,000.$$

So the initial value is \$16,000 and represents the purchase price of the car. The model is given over a 15-year period of time. In this context, we can consider both variables, t and $V(t)$ to be continuous. So the real-world domain is the interval

$$\text{Domain} = [0 \text{ yrs., } 15 \text{ yrs.}].$$

The corresponding range is the interval of output values for the function $V(t)$ with the largest value at $t = 0$ and the lowest value at $t = 15$ (decay):

$$\text{Range} = [\$2351.58, \$16000].$$

The graph joins the y-intercept $(0, 16000)$ and the point $(15, 2351.58)$ by a solid line. The asymptote has no good interpretation outside the real-world domain.

Example 3.9

The amount in grams of a sample of radioactive substance after t decades is given by the function $A(t) = 10,000 \cdot \left(\frac{5}{4}\right)^{-t}$.

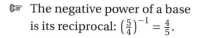 The negative power of a base is its reciprocal: $\left(\frac{5}{4}\right)^{-1} = \frac{4}{5}$.

Solution: We rewrite the equation using the property of negative exponents:

$$A(t) = 10,000 \cdot \left(\frac{5}{4}\right)^{-t} = 10,000 \cdot \left(\frac{4}{5}\right)^{t} = 10,000 \cdot (0.8)^{t}.$$

The base of the exponential is $b = 0.8 < 1$, so the mass $A(t)$ of the sample is an exponential decay function with the decay factor of 0.8. The initial value is

$$A(0) = 10,000 \cdot (0.8)^{0} = 10,000 \cdot 1 = 10,000 \text{ grams}.$$

The real world domain is the interval of all positive real numbers including zero

Domain = $[0, +\infty)$ decades   OR   Domain = the interval $t \geq 0$ decades.

The corresponding range is the interval from 0 to 10,000 grams written as

Range = $(0, 10000]$ grams   OR   Range = the interval $0 < A(t) \leq 10,000$ grams.

The graph is sketched in the margin with the y-intercept $(0, a) = (0, 10000)$ and the x-axis as an asymptote to the right.

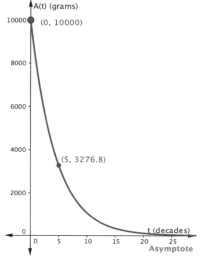

Figure 3.9 Radioactive Decay

Wrapping Up

In this section we learned about the two types of exponential functions - growth and decay. Exponential functions can be written in the form $y = f(x) = a \cdot b^{x}$, and the base b, which is also the constant multiplier when inputs increase by 1, determines whether or not the function grows or decays. If b is between 0 and 1 the function represents decay, if b is greater than 1 it represents growth.

Before making a graph, we find the real-world domain and range so we can choose scales that show the overall pattern (or trend) of the function. Scales may need to handle very large numbers which can distort the picture. In this case it is helpful to create a zoomed-in version to show the initial value. Exponential growth graphs increase while decay graphs decrease, but their general shapes are just reflections of each other across the y-axis. Like all linear functions, exponential functions have one y-intercept $(0, a)$, where a is the initial value of the function. But exponential graphs are different from linear graphs in that they do not have x-intercepts. Instead they are asymptotic - they get closer and closer to the x-axis without ever touching or crossing it.

In the next section we will answer questions and solve problems about more real-world situations that involve exponential growth and decay.

Exercises

Exercises for 3.2 Features of Exponential Function Graphs

P3.13   Explain what makes a function an exponential. Refer to equations, graphs, and tables. Give some examples of exponential functions and some examples of functions that are not exponential and explain why they aren't exponential.

P3.14   **a)** Describe the shape of the graph of an exponential function. Then sketch some (small) examples on paper.

   b) Explain how you determine whether an exponential function is increasing (growth) or decreasing (decay). Refer to equations, graphs, and tables.

   c) Can an exponential function graph have x-intercepts? Explain the meaning of a horizontal asymptote.

P3.15   For each of the following situations

   • Decide if the function represents growth or decay and explain.

   • Identify the growth/decay factor and the initial value.

   • Determine the real-world domain and range.

   • Sketch a graph, label the y-intercept and the asymptote.

   a) The deer population in one area of New England is given by the function $P(t) = 12,000(1.08)^t$ for t years since 1990 and this pattern continued until 2010.

   b) The value of a new car is given by $V(t) = \$19,000(0.8)^t$, t years after purchase, and this trend will continue for 15 years.

   c) For the next 30 years the value of a stamp collection is given by the function $S(t) = \$8,400(1.035)^t$, t years after purchase.

   d) The value of a house is given by the function $H(t) = \$220,00(1.021)^t$, t years after its purchase, for the next 20 years.

   e) The number of cells of a strain of bacteria in a sample is modeled by the function $B(d) = 3 \times 10^6 (0.75)^d$ after d days.

P3.16   Identify which functions are exponential and which are not and explain why. For those that *are* exponential,

   • Determine whether the function represents growth or decay and explain why.

   • Identify the growth/decay factor and the initial value.

   • Sketch a graph of the function, label the y-intercept and the asymptote.

   a) $y = 64 \cdot (1.25)^x$

   b) $A(t) = 27,000 \cdot \left(\frac{1}{3}\right)^t$

   c) $N(x) = 1,000,000 \cdot (0.6)^x$

   d) $A(t) = 100,000 \cdot \left(\frac{5}{4}\right)^{-t}$

   e) $y = 10 \cdot (0.35x)^2$

   f) $N(x) = 4000 \cdot (1.045)^x$

   g) $f(t) = 1600 \cdot (4)^{-t}$

   h) $y = 20 \cdot (-1.4)^x$

   i) $y = 0.4 \cdot (2)^{3t}$ (optional)

   j) $y = 0.5 \cdot (4)^{t/2}$ (optional)

P3.17   Write an example of an equation for a function, if any, that satisfies each of the following:

   a) An exponential growth function with an initial value of 100.

   b) An exponential decay function with an initial value of 4000.

   c) An exponential growth function with an initial value of 0.6.

   d) An increasing linear (growth) function with an initial value of 60.

   e) A decreasing linear (decay) function with an initial value of 2.

   f) An exponential function with a base of 1.8.

   g) An exponential function with a constant multiplier of 1/3.

   h) An exponential function with a growth factor of 2.1.

   i) An exponential function with a decay factor of 0.1.

   j) An exponential function with a decay factor of −3. (optional)

P3.18 From the equation of the function, determine whether the graph is exponential or linear and whether it is increasing or decreasing. Then find the intercept(s). Use that information to choose the graph in **Fig 3.10** below that matches the function.

a) $f(x) = -6x - 9$

b) $y = 9x - 8$

c) $f(x) = 12(6)^x$

d) $y = 2(7/8)^x$

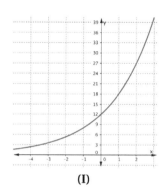
(I)

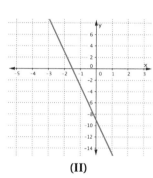
(II)

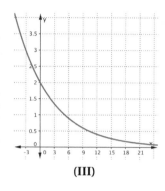
(III)

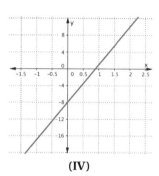
(IV)

Figure 3.10

Exercises for 3.2 Features of Exponential Function Graphs

P3.19 For each of the following tables do the following tasks:

- Does the table represent an exponential function? Explain why or why not.

- IF it DOES represent an exponential function, determine whether the function represents growth or decay and explain why.

- Write an equation for the function and sketch its graph.

a)

d days	$N(d)$ cells
0	24,000
1	30,000
2	37,500
3	46,875

b)

d days	$N(d)$ cells
0	45,000
1	30,000
2	20,000
3	13,333

c)

t years	$P(t)$ people
0	220,000
1	231,000
2	242,550
3	254,677.5

d)

t years	$P(t)$ people
0	28,000
1	21,000
2	14,000
3	7,000

e)

d days	$N(d)$ cells
0	24,000
1	18,000
2	13,500
3	10,125

f)

d days	$N(d)$ cells
2	81000
3	54000
4	36000

g)

d days	$N(d)$ cells
3	24000
4	36000
5	54000

i)

x	$F(x)$
8	8,192
9	32,768
10	131,072

h)

x	y
4	200
5	600
6	1800

P3.20 Given each graph in **Fig 3.11** determine whether it is linear or exponential and write its equation using the labeled points.

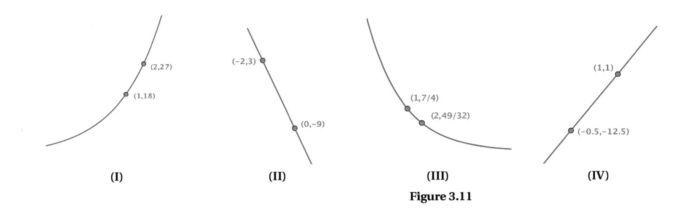

(I)   (II)   (III)   (IV)

Figure 3.11

3.3 Problem Solving with Exponentials

Launch Exploration

Let us buy a house for $100,000. Do real estate values increase or decrease? Explain. Suppose that in our town the value of houses increase by 2% every year.

- What are the variables and the constants in this situation? Identify the independent and dependent variables including units.

- How do you calculate the value of the house after the first year? After the second year? What does a 2% yearly increase mean? Enter values in a table.

- Is this a linear relationship? Why or why not? Is it exponential? Explain.

- What is the added value, in dollars, of the house after 1 year? After 2 years?

Financial Note: Banks and other financial institutions calculate rates in different ways - compounding annually, monthly, daily, etc. When we say that the value of the house increases by 2% every year with no extra information, we assume that 2% represents the *effective* rate of increase per year, which means that the compounding details have already been worked in.

Relative Rate of Growth

In real estate, land increases in value and the structures decrease in value unless the owners invest in maintaining their condition and value. In the **Launch Exploration** we describe a long term real estate model over which values generally increase overall.

> **Buying a House**
>
> What are the variables?
>
> DV: $V(t)$ = value of house ($)
> IV: t = time the house is owned (yrs)   ☞ $100\% = \frac{100}{100} = 1$
>
> The rate $2\% = \frac{2}{100} = 0.02$ means $2 added value to every $100 existing value. The units are $ per $ and this is a *relative* rate of increase in value compared to the existing value. To get the total value of the house after the first year we add to the original value of $100,000 the *extra* value of **2%** of $100,000 and get
>
> $$\text{1st year value} = \mathbf{100\%} \text{ of } \$100,000 + \mathbf{2\%} \text{ of } \$100,000$$
> $$= \mathbf{1}(\$100000) + (\mathbf{.02})(\$100000)$$
> $$= (\mathbf{1 + .02})(\$100000) \qquad \text{(Common factor)}$$
> $$= (\mathbf{1.02})(\$100000) = \$102,000.$$

In the second year we earn 2% of the first year value. To find the extra value we take 2% of $102,000 not of the initial value $100,000. So we multiply $(.02)(\$102000)$ and add this value to the last amount

$$\text{2nd year value} = \mathbf{100\%} \text{ of } \$102,000 + \mathbf{2\%} \text{ of } \$102,000$$

$$= 1(\$102000) + (.02)(\$102000)$$

$$= (1 + .02)(\$102000)$$

$$= (1.02)(\$102000) = \$104,040.$$

Diff	t	$V(t)$	Diff
+1	0	100,000	+2000
+1	1	102,000	+2040
	2	104,040	

Table 3.13 Linear? - No

(Constant multiplier)

The current year amount is the previous year amount times a *constant multiplier*.

$$\text{current amount} = \mathbf{100\%}(\ ) + \mathbf{2\%}(\ )$$

$$= (\mathbf{1} + \mathbf{.02})(\ )$$

$$= (\mathbf{1.02})(\text{previous amount}).$$

This is an exponential function (See **Table 3.14**)

$$(\mathbf{3.9}) \qquad\qquad V(t) = 100,000(\mathbf{1.02})^{t}$$

with the initial value $a = \$100,000$ and the constant *growth factor* $b = \mathbf{1.02}$. This growth factor tells us that each year the value of the house is **102%** of the previous year's value where **2%** is the percent of added value.

Diff	t	$V(t)$	Mult
+1	0	100,000	×1.02
+1	1	102,000	×1.02
	2	104,040	

Table 3.14 Constant multiplier/ factor (Exponential? - Yes)

In general, the fraction or percentage of the output that is *gained* when the input increases by 1 unit is called the *relative rate of growth* such that

$$(\mathbf{3.10}) \qquad\qquad \textbf{growth factor} = 1 + \textbf{growth rate}.$$

The growth factor is the fraction or percentage of the previous output that remains to give the current output, including the added value.

Definition 3.7 *Relative Growth Rate*

The **(relative) growth rate** r of an exponential function $y = a \cdot b^{x}$ is the fraction or percentage of the output that is *gained* when the input increases by 1 unit with the **growth factor** $b > 1$ such that $b = 1 + r$.

Relative Rate of Decay

Buying a Car

The value of a certain new car that costs $24,500 decreases in value by 7.8% each year after its purchase for the first 10 years.

What are the variables? The independent variable is t = time the car is owned (yrs) and the dependent variable is $V(t)$ = value of car ($).

The rate $7.8\% = \frac{7.8}{100} = 0.922$ means \$7.8 lost value to every \$100 existing value. The units are \$ per \$ and this is a *relative* rate of decrease in value compared to the existing value. To get the total value of the car after the first year we subtract from the original value of \$24,500 the *lost* value of **7.8%** of \$24,500 and get

$$\text{1st year value} = \textbf{100\%} \text{ of } \$24,500 - \textbf{7.8\%} \text{ of } \$24,500$$
$$= \textbf{1}(\$24500) - (\textbf{.078})(\$24500)$$
$$= (\textbf{1} - \textbf{.078})(\$24500) \qquad \text{(Common factor)}$$
$$= (\textbf{0.922})(\$24500) = \$22,589.$$

In the second year we lose 7.8% of the first year value. To find the lost value we take 7.8% of \$22,589 not of the initial value \$24,5000. So we multiply $(\textbf{.078})(\$24500)$ and subtract this value from the last amount

$$\text{2nd year value} = \textbf{100\%} \text{ of } \$22,589 - \textbf{7.8\%} \text{ of } \$22,5890$$
$$= \textbf{1}(\$22589) - (\textbf{.078})(\$22589)$$
$$= (\textbf{1} - \textbf{.078})(\$22589)$$
$$= (\textbf{0.922})(\$22589) = \$20,827.10.$$

Diff	t	V(t)	Diff
+1	0	24,500.00	−1911
+1	1	22,589.00	−1716.9
	2	20,872.10	

Table 3.15 Linear? - No

The current year amount is the previous year amount times a *constant multiplier*.

$$\text{current amount} = \textbf{100\%}(\ ) - \textbf{7.8\%}(\ )$$
$$= (\textbf{1} - \textbf{.078})(\ )$$
$$= (\textbf{0.922})(\text{previous amount}). \qquad \text{(Constant multiplier)}$$

This is an exponential function (See **Table 3.16**)

(**3.11**) $$V(t) = 24,500(\textbf{0.922})^{t}$$

with the initial value $a = \$24,500$ and the constant *decay factor* $b = \textbf{0.922}$. This decay factor tells us that each year the value of the car is **92.2%** of the previous year's value where **7.8%** is the percent of lost value.

Diff	t	V(t)	Mult
+1	0	24,500.00	×0.922
+1	1	22,589.00	×0.922
	2	20,872.10	

Table 3.16 Constant multiplier/factor (Exponential? - Yes)

In general, the fraction or percentage of the output that is *lost* when the input increases by 1 unit is called the *relative rate of decay* such that

(**3.12**) $$\textbf{decay factor} = \textbf{1} - \textbf{decay rate}.$$

The decay factor is the fraction or percentage of the previous output that remains to give the current output, excluding the lost value.

Definition 3.8 *Relative Decay Rate*

The **(relative) decay rate** r of an exponential function $y = a \cdot b^{x}$ is the fraction or percentage of the output that is *lost* when the input increases by 1 unit with the **decay factor** $0 < b < 1$ such that $b = 1 - r$.

Exponential Growth/Decay Rates in Function Equations

For each function, determine whether or not it is exponential and explain why.

IF it is exponential, identify the initial value, the growth/decay factor, and the growth/decay rate.

a) $y = 64(1.24)^x$ is exponential with initial value $a = 64$, growth factor $b = 1.24 = 124\%$ which means each output is 124% of the previous output, and growth rate

$$r = 1.24 - 1 = 0.24 = 24\%$$

which means the added value is 24% of the previous output.

b) $A(t) = 28,000(1/4)^t$ is exponential with the initial value $a = 28,000$, the decay factor $b = 1/4$ which means each output is $1/4$ of the previous output, and the decay rate

$$r = 1 - 1/4 = 3/4$$

which means the lost value is $3/4$ of the previous output.

c) $f(x) = 21(-1.03)^x$ is not well defined as the base -1.03 cannot be negative.

d) $P(t) = 140,000(0.97)^t$ is exponential with the initial value $a = 140,000$, the decay factor $b = 0.97 = 97\%$ (each output is 97% of the previous output), and the decay rate $r = 1 - 0.97 = 0.03 = 3\%$ (the lost value is 3% of the previous output).

☞ $\left(\frac{3}{5}\right)^{-t} = \left(\frac{5}{3}\right)^{t}$

e) $V(t) = 9500(3/5)^{-t} = 9500(5/3)^t$ is exponential with initial value $a = 9500$, *growth* factor $b = 5/3$, and *growth* rate $r = 5/3 - 1 = 2/3$.

Exponential Equation - Another Example of Nonlinear Equation

An **exponential equation** is an equation of the form

(3.13) $b^x = N,$ $(b > 0$ and $b \neq 1).$

where the exponent x is a variable and the base b is a *positive* real number $\neq 1$. If N is a *positive* real number $(N > 0)$, then this equation has a *unique* real solution denoted by $x = \log_b N$ and called *the logarithm in base b of N*.

☞ $(1.02)^t = 2$

can be written as

$t = \log_{1.02}(2)$

(3.14) $(\text{base})^{\text{exponent}} = \text{N}$ means $\text{exponent} = \log_{\text{base}}(\text{N}).$

Definition 3.9 *Logarithm of a Number*

The **logarithm in base b of a positive real number** N is denoted by $\log_b N$ and is the real number *solution* to the equation $b^x = N$, i.e.

$x = \log_b N$ is the *exponent* in $b^x = N$ $(b > 0, \ b \neq 1),$

If N is *nonpositive* $(N \leq 0)$, the **Equation (3.13)** has no real solutions because the exponential b^x is always positive and does not take on negative or zero values.

Exponential vs. Power Equations

☞ On a calculator,
$\log(a) = \log_{10}(a)$ (common)
$\ln(a) = \log_e(a)$ (natural)
$e \approx 2.718...$ (Euler's number)
$\log_b(a) = \frac{\log(a)}{\log(b)} = \frac{\ln(a)}{\ln(b)}$

In power equations we solve for the base and use roots. In exponential equations we solve for the exponent and use logarithms.

a) $(1.24)^x = 800$ means

$$x = \log_{1.24}(800) = \frac{\log(800)}{\log(1.24)} \approx 31.08.$$

b) $140,000(0.97)^t = 70,000$ or $(0.97)^t = 1/2$ which means

$$t = \log_{0.97}(1/2) = \frac{\ln(1/2)}{\ln(0.97)} \approx 22.76.$$

c) $200,000 = 400,000(5/3)^x$ or $1/2 = (5/3)^x$ which means

$$x = \log_{5/3}(1/2) = \frac{\log(1/2)}{\log(5/3)} \approx -1.36.$$

d) $21,000 = 21x^2$ or $1,000 = x^2$ so that $x = \sqrt[2]{1000} \approx 31.6$ or $x = -\sqrt[2]{1000} \approx -31.6$.

e) $x^2 + 36 = 0$ or $x^2 = -36$ has no solutions since a square cannot be negative.

f) $2^x = -8$ has no solutions since an exponential cannot be negative.

g) $0 = 9500(7/4)^{-t}$ has no solutions since an exponential cannot be zero.

h) $-21,000 = 21x^3$ or $-1,000 = x^3$ so that $x = \sqrt[3]{-1000} = -10$.

(a)	**d** \# days	**N(d)** \# cells after d days
	0	24,000
	1	30,000
	2	37,500
	3	46,875
	10	
		682,121

(b)	**d** \# days	**N(d)** \# cells after d days
	4	20,736
	5	15,552
	6	11,664
	7	8,748
	9	
		657

Figure 3.12 Exponential Tables

More Exponential Modeling

For each example in this section we consider the variables to be *continuous*.

Extending exponential tables

For each table in **Fig 3.12**:

- Verify that the function is exponential. Explain.
- Identify the initial value, the growth/decay factor, and the growth/decay rate.
- Write an equation for the function.
- Use the equation to fill in the missing values in the table.

a) Table 3.17 is exponential since the successive output ratios are constant as the inputs increase by 1. The constant growth factor is

$$b = \frac{30,000}{24,000} = \frac{37,500}{30,000} = \frac{46,875}{37,500} = 1.25 > 1.$$

Diff	**d**	**N(d)**	Mult
+1	0	24,000	⟩ ×1.25
+1	1	30,000	⟩ ×1.25
+1	2	37,500	⟩ ×1.25
	3	46,875	
	10	223,516	
	15	682,121	

Table 3.17 Exponential growth

The growth rate is $r = 1.25 - 1 = 0.25 = 25\%$ per day. The initial value is $a = 24,000$ cells when the input equals zero ($d = 0$) in the table. So the function equation is

$$N(d) = a \cdot b^d = 24,000(1.25)^d.$$

The missing value when the input $d = 10$ is

$$N(10) = 24,000(1.25)^{10} \approx 223,516 \text{ cells.}$$

To find the input when output is $N(d) = 682,121 = 24,000(1.25)^d$, we solve for d:

$$(1.25)^d = \frac{682,121}{24,000} \approx 28.42, \qquad d = \log_{1.25}(28.42) = \frac{\log(28.42)}{\log(1.25)} \approx 15 \text{ days.}$$

b) **Table 3.18** is exponential since the successive output ratios are constant as the inputs increase by 1. The constant decay factor is

$$b = \frac{15,552}{20,736} = \frac{11,664}{15,552} = \frac{8,748}{11,664} = 0.75 < 1.$$

The decay rate is $r = 1 - 0.75 = 0.25 = 25\%$ per day. To find the initial value we pick an ordered pair solution from the table, say $(7, 8748)$, and substitute this pair and the factor $b = 0.75$ in the function equation:

$$N(d) = a \cdot b^d, \qquad 8,748 = a(0.75)^7, \qquad a = \frac{8,748}{(0.75)^7} = 65,536 \text{ cells.}$$

So the function equation is

$$N(d) = a \cdot b^d = 65,536(0.75)^d.$$

The missing value when the input $d = 9$ is

$$N(9) = 65,536(0.75)^9 \approx 4,921 \text{ cells.}$$

To find the input when output is $N(d) = 657 = 65,536(0.75)^d$, we solve for d:

$$(0.75)^d = \frac{657}{65,536} \approx 0.01, \qquad d = \log_{0.75}(0.01) = \frac{\log(0.01)}{\log(0.75)} \approx 16 \text{ days.}$$

Diff	d	$N(d)$	Mult
+1	4	20,736	×0.75
+1	5	15,552	×0.75
+1	6	11,664	×0.75
	7	8,748	
	9	4,921	
	16	657	

Table 3.18 Exponential decay

Example 3.10 The value of a house in the next 40 years after purchase is given by the **Equation (3.9)**. Find the real world domain and range for this model and answer the following questions:

a) How much is the house worth after 9 months?

b) How long will it take for the value of the house to double?

c) You plan to sell the house when it reaches $250,000 in value. According to this model, how many years will it take for the house to reach that value?

Solution: The value of the house in dollars is $V(t) = \$100,000(1.02)^t$ where t is the time the house is owned in years. From the verbal description, the real-world domain is the interval [0 yrs, 40 yrs]. The corresponding range is from the initial value $V(0) = \$100,000$ to

$$V(20) = \$100,000(1.02)^{40} = \$220,803.97.$$

In interval notion, the range is [$100000, $220,803.97].

a) We convert 9 months in 9/12= 0.75 years so the house is worth

$$V(0.75) = \$100,000(1.02)^{0.75} \approx \$101,496.28.$$

☞ The time when the amount is double the initial amount is called the **doubling time**.

b) We solve the equation $V(t) = (2)(100000) = 100,000(1.02)^t$ for t:

$$2 = (1.02)^t, \qquad t = \log_{1.02}(2) = \frac{\log(2)}{\log(1.02)} \approx 35 \text{ years.}$$

Since the input 35 yrs is in the domain [0 yrs, 40 yrs], the answer is valid.

c) We solve the equation $250,000 = 100,000(1.02)^t$ for t:

$$2.5 = (1.02)^t, \qquad t = \log_{1.02}(2.5) = \frac{\log(2.5)}{\log(1.02)} \approx 46 \text{ years.}$$

However, the input 46 yrs is not in the domain [0 yrs, 40 yrs], so it is *invalid* for this model. We can see that *without* calculations since the output $250,000 is not in the range [$100000, $220803.97] of this model either.

Example 3.11 The value of a car in the next 10 years after purchase is given by the **Equation (3.11)**. Find the real world domain and range for this model and answer the following questions:

a) What is the car worth 39 months after purchase?

b) How long will it take for the car to be worth half of its original value?

c) When will the car be worth only $600?

Solution: The value of the car in dollars is $V(t) = \$24,500(0.922)^t$ where t is the time the car is owned in years. From the verbal description, the real-world domain is the interval [0 yrs, 10 yrs]. The corresponding range is from the initial value $V(0) = \$24,500$ to

$$V(10) = \$24,500(0.922)^{10} = \$10,876.15.$$

In interval notation, the range is [$10876.15, $24500].

a) We convert 39 months into $39/12 = 3.25$ years so the car is worth

$$V(3.25) = \$24,500(0.922)^{3.25} \approx \$18,816.62.$$

☞ The time when the amount is half the initial amount is called the **half-life**.

Keys to Exponential Modeling

Identify (choose) the independent and dependent variables.

Verify the relation is exponential (verbal, equation, table, graph).

Find the constant growth/decay factor and the initial value.

Write an exponential equation and give a real-world domain & range.

Table 3.19

IV (In)	DV (Out)
t (hrs)	$M(t)$ (mg)
0	500
4	379.4
17	154.7
?	150
18	144.4

Table 3.20 $M(t) = 500\left(\frac{14}{15}\right)^t$.

---

b) We solve the equation $V(t) = \frac{1}{2}(24500) = 24{,}500(0.922)^t$ for t:

$$\frac{1}{2} = (0.922)^t, \qquad t = \log_{0.922}\left(\tfrac{1}{2}\right) = \frac{\log(0.5)}{\log(0.922)} \approx 8.5 \text{ years.}$$

Since the input 8.5 yrs is in the domain [0 yrs, 10 yrs], the answer is valid.

c) We solve the equation $600 = 24{,}500(0.922)^t$ for t:

$$\frac{6}{245} = \frac{600}{24{,}500} = (0.922)^t, \quad t = \log_{0.922}\left(\frac{6}{245}\right) = \frac{\log(6/245)}{\log(0.922)} \approx 46 \text{ years.}$$

However, the input 46 yrs is not in the domain [0 yrs, 10 yrs], so the answer is *invalid* for this model. We can see that *without* calculations since the output \$600 is not in the range [\$10876.15, \$24500] of this model either.

Example 3.12 A doctor administers 500 mg of medicine to a patient. The level of the drug in the bloodstream decreases by $1/15$ every hour. Do the tasks in **Table 3.19** and answer the following questions:

a) How much medicine is in the bloodstream after 4 hours?

b) If the patient needs at least 150 mg in her blood, after how many hours should she get another dose?

Solution: The independent variable is time t in hours and the dependent variable is the amount of medicine $M(t)$ in mg. The decay rate is $r = 1/15$ per hour so that the decay factor is $b = 1 - 1/15 = 14/15$. The initial value is $a = 500$ mg. So the function equation is

$$M(t) = a \cdot b^t = 500\left(\frac{14}{15}\right)^t.$$

To find the domain we need to find the time when the patient is drug free. For example that could be when the medicine is under 1 mg:

$$M(t) = 1 = 500\left(\frac{14}{15}\right)^t, \quad \left(\frac{14}{15}\right)^t = \frac{1}{500}, \quad t = \log_{14/15}(1/500) \approx 90 \text{ hours.}$$

So the domain is the time interval $[0, 90]$ hours and the range is $[1, 500]$ mg.

a) After 4 hours we have $M(4) = 500\left(\frac{14}{15}\right)^4 \approx 379.4$ mg left in bloodstream.

b) We solve the equation $M(t) = 150 = 500\left(\frac{14}{15}\right)^t$ for t:

$$\frac{150}{500} = \left(\frac{14}{15}\right)^t, \qquad t = \log_{14/15}\left(\frac{150}{500}\right) = \frac{\log(150/500)}{\log(14/15)} \approx 17.5 \text{ hours.}$$

The value $t = 17.5$ can be "interpolated" from **Table 3.20** by finding two values of the function such that 150 is between the two values.

Wrapping Up

Exponential models are used in finance and natural sciences and in this section we have seen how to create them by using the relative rate of growth or decay and finding the constant multiplier. This relative rate expresses the fraction or percentage of the output that is gained or lost when the input increases by 1 unit. When the constant multiplier b is known, the relative rate r is the positive solution to the equation $b = 1 \pm r$.

The initial value is then found by using an ordered pair solution from tables or situations and the function equation then can be used to answer questions. When we know the input the output is obtained by simple evaluation. When we know the output, the input is obtained by solving an exponential equation of the form $b^x = N$ for the exponent x. For estimates we can use exponential tables of values. For an exact value, we can use the log operation $x = \log_b(N)$ and a calculator with the fast keys $\log = \log_{10}$ (common log) and $\ln = \log_e$ (natural log) giving $\log_b(N) = \log(N) / \log(b)$ or $\ln(N) / \ln(b)$. This property of logs can be derived by simple manipulations (not included). Now you learned the algebra you need to understand exponential models and their many applications.

Exercises

Exercises for 3.3 Problem Solving with Exponentials

P3.21 Write an equation for exponential growth with an initial value of 400,000 and a growth rate of 4.5%. What is the growth factor for your function? Sketch a graph for your function. Label any important points.

P3.22 Write an equation for exponential decay with an initial value of 800 and a decay rate of 1/9. What is the decay factor for your function? Sketch a graph for your function. Label any important points.

For the situations in each example:

- Identify and label the variables, including which is independent and which is dependent, and the constants in the situation.

- Write an equation that models the situation.

- Use the function equation to answer the following questions. Express each answer as an ordered pair and in function notation. Enter each of your answers in a table.

P3.23 Between 1985 and 2015 the world's jaguar population decreased at a rate of about 2.2%. In 1985 there were an estimated 25,000 jaguars.

- **a)** About how many jaguars were there in 1990? In 1995? Did we lose the same number of jaguars in each of those 5-year time periods? Explain.

- **b)** In what year did the population reach 16,750?

- **c)** Give the domain and range for the model.

- **d)** According to the model, what will the jaguar population be in 2030? Do you think this is an accurate prediction? Explain.

P3.24 A colony of bacteria begins with 100 cells and grows exponentially at a rate of 1/9 every day.

- **a)** About how many cells are there in the colony after two weeks?

- **b)** How long will it take until there are 1,000,000 cells?

P3.25 There is an outbreak of the Ebola virus. There are 12 cases and the cases will increase by 7% each day for two months without immunization.

- **a)** Without immunization, how many cases will there be after one week?

- **b)** Will the number of cases reach 1200? Explain.

- **c)** What are the real-world domain and range for this function?

P3.26 In 1859, a southern Australian farmer, homesick for England, imported two dozen (24) wild English rabbits and set them free on his land. For the next six years the rabbit population grew by about 21% each month.

- **a)** How many rabbits were there after 5 months?

- **b)** How many rabbits were there after one year? Two years?

- **c)** Did the population reach a quarter of a million rabbits? If so, after how long in months? In years?

- **d)** How many rabbits were there at the end of the 6-year period?

- **e)** Do you think the model remained valid past the 6 years? Explain. (And you can look it up!)

P3.27 You purchase a house for $220,000 and the realtor says that the home values in that area increase by 2.1% each year.

- **a)** How much is the house worth after 9 months?

- **b)** How much is the house worth after 4 years?

- **c)** You plan to sell the house when it reaches $350,000 in value. According to this model, how many years will that take?

P3.28 The value of a certain new car that costs $32,500 decreases in value by 7.8% each year after its purchase for the first 10 years.

- **a)** How much time will it take for the car to be worth half of its original value?

b) What is the car worth a year and a half after purchase?

c) What is the car worth 3 years after purchase?

P3.29 A strain of bacteria starts with 10,000,000 cells. Every hour 2/7 of the cells die.

a) How many cells remain after 15 minutes? After 3 hours?

b) Estimate how long it will take for there to be $\frac{1}{2}$ of the original number of cells left.

c) Estimate how long it will take for there to be $\frac{1}{4}$ of the original number of cells left.

d) Give the real-world domain and range for this model.

P3.30 A doctor administers 500 mg of medicine to a patient. The level of the drug in the bloodstream decreases by 6% every hour.

a) How much medicine is in the bloodstream after 4 hours?

b) If the patient needs at least 150 mg in her bloodstream, after how many hours should she get another dose?

P3.31 A strain of virus in a host starts with 20 cells. Every day there are 6/5 the number of cells from the day before. The virus continues to get reproduced like this until the host gets medication one month later.

a) How many cells are there after 6 days? After two weeks?

b) Estimate how long it will take for there to be twice the original number of cells.

c) Estimate how long it will take for there to be 1,100 cells.

d) Give the real-world domain and range for this model.

P3.32 A strain of bacteria in a petri dish starts with 200,000,000 cells. Every day 1/3 of the cells die.

a) How many cells remain after 6 days? After two weeks?

b) Estimate how long it will take for there to be $\frac{1}{2}$ of the original number of cells left.

c) Estimate how long it will take for there to be 60,000,000 cells left.

d) Give the real-world domain and range for this model.

P3.33 For each table determine whether or not it is an exponential function. Explain why or why not. IF the table is exponential:

• Identify the initial value, the growth/decay factor, and the growth/decay rate.

• Write an equation for the function.

• Use the equation to fill in the missing values in the table.

a)

d days	$N(d)$ cells
0	24,000
1	30,000
2	37,500
3	46,875
10	
	682,121

b)

d days	$N(d)$ cells
0	45,000
1	30,000
2	20,000
3	13,333
8	
	780.37

c)

t years	$P(t)$ people
0	220,000
1	231,000
2	242,550
3	254,677.5
5	
	309,562

d)

t years	$P(t)$ people
2	2,250,000
3	2,700,000
4	3,240,000
6	
	5,598,720

e)

d days	$N(d)$ cells
4	20,736
5	15,552
6	11,664
7	8,748
9	
	657

P3.34 For each equation determine whether or not it is exponential and explain why. IF it is exponential, identify the initial value, the growth/decay factor, and the growth/decay rate.

a) $y = 64(1.24)^x$

b) $A(t) = 28,000 \left(\frac{1}{4}\right)^t$

c) $N(x) = 200,000 \left(\frac{5}{3}\right)^x$

d) $f(x) = 21(-1.03)^x$

e) $P(t) = 140,000(0.97)^t$

f) $V(t) = 9500 \left(\frac{5}{3}\right)^{-t}$

P3.35 Solve the following equations:

a) $800 = 64(1.24)^x$

b) $10,000 = 28,000 \left(\frac{1}{4}\right)^t$

c) $400,000 = 200,000 \left(\frac{5}{3}\right)^x$

d) $21,000 = 21x^3$

e) $70,000 = 140,000(0.97)^t$

f) $19,000 = 9500 \left(\frac{5}{3}\right)^{-t}$

g) $264 = 64 - 4x + 2$

h) $21,000 = 21x^2$

i) $12 = 1.5(1.051)^x$

j) $400(0.968)^x = 360$

k) $19x + 42 - 4x = 161 + x$

l) $x^2 + 25 = 0$

Unit 4

Quadratic Models

A Review of Main Ideas

Before we begin this unit, summarize the characteristics of *linear* and *exponential* functions. For each type of function,

- Explain how you identify each kind of relationship in situations, tables, graphs, and equations.

- How do you know when the function is increasing/decreasing?

- Does the function have x- and y-intercepts? How many can they have?

- Give examples and illustrate using each representation.

4.1 Another New Pattern - Introduction to Quadratic Functions

Launch Exploration

A New Example: Suppose we are planning a square-shaped garden. Let

IV (Input): s = the side length of a square garden
DV (Output): $A(s)$ = the area of the square garden

We must choose appropriate units for each variable. Draw such a garden for some different side lengths. Get together with a group of classmates and do the tasks in **Table 4.1** on the margin.

Quadratic Function Equations

Quadratic function *equations* (*formulas*) can always be written in the form:

$$y = f(x) = ax^2 + bx + c, \qquad \text{where } a \neq 0. \qquad \text{(Why } a \neq 0\text{ ?)}$$

Keys to Modeling

Identify the dependent vs. independent variables including units.

Make a table of values.

Examine the table for a pattern.

Linear? Exponential? Neither? EXPLAIN.

Identify the real-world domain and range.

Create a graph for the function.

Write an equation (formula) to represent the function.

Table 4.1

Quadratic functions are given by *polynomials* where the highest power of the (input) variable is 2. In other words, there is a squared (quadratic) *term* and no higher degree term. Here a, b, and c refer to the numerical *coefficients* of the quadratic (x squared), linear (x to the first), and constant terms respectively.[4]

☞ If $a = 0$, then the function would just be linear, but $b = 0$ or $c = 0$ are allowed.

Definition 4.1 *Quadratic Function Equation (Formula)*

A **quadratic function** can be defined by an equation (formula) of the form

$$y = f(x) = ax^2 + bx + c, \ a \neq 0,$$

where x is the independent variable (input), y is the dependent variable (output), and a, b, c are constants.

Examples of Quadratic Functions

The following equations (formulas) define quadratic functions:

$y = x^2$	$a = 1,$	$b = 0,$	$c = 0$
$f(x) = 5x^2 + 3 - 2x$	$a = 5,$	$b = -2,$	$c = 3$
$h(s) = 1000 - 40s^2$	$a = -40,$	$b = 0,$	$c = 1000$
$y = -4.9t^2 - 25t$	$a = -4.9,$	$b = -25,$	$c = 0.$

☞ Recall that a goes with the quadratic term, b with the linear term, and c with the constant term.

Non-Examples of Quadratic Functions

The following equations (formulas) do not define quadratic functions:

$y = 7x^2 + 3x^3 - 8$	Not quadratic because there is a cubic term ($3x^3$). This is a cubic function.
$V(r) = 3r - 8$	Not quadratic because there is no quadratic (squared) term. (What kind of function is this?)
$y = 20(1.2)^t$	Not quadratic because it is not a polynomial function. (What kind of function is this?)

Quadratic Function Tables

Going back to planning a square-shape garden, we must choose units for each variable such that they agree with each other in formulas:

☞ If we choose meter for side-length we should choose square meters for area.

IV (Input): s = the side length of a square garden (feet)
DV (Output): $A(s)$ = the area of the square garden (square feet)

Below we draw some examples of square-shape gardens for side lengths $s = 1$, 2, 3, 4, 5 feet. Then we make a **table of values** for the area in square feet by counting the unit squares enclosed inside each garden.

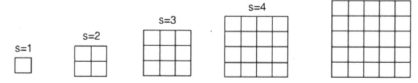

Figure 4.1 Square-shape gardens.

Diff	s	$A(s)$	1st Diff
+1	1	1	+3
+1	2	4	+5
+1	3	9	+7
+1	4	16	+9
	5	25	

Table 4.2 Linear? - No

Next let us examine the pattern in the table to check if it is linear or exponential. *When inputs increase by one in the table of values, linear functions have constant first differences between outputs*, which is the constant rate of change or slope. We can see in **Table 4.2** that $A(s)$ is clearly not linear. *When inputs increase by a constant, the outputs in an exponential function change by a constant multiplier* (factor.) We can see in **Table 4.3** that $A(s)$ is not exponential. Although, in **Table 4.2**, the first differences are not constant, they increase by a constant of 2 (see **Table 4.4**), so $A(s)$ has *constant second differences*. This is a new pattern.

☞ The first differences in a linear pattern are not the slope unless the inputs increase by a constant of one (slope = $\Delta y / \Delta x$.)

Definition 4.2 *Second Order Differences*

In a table of values of a function $y = f(x)$, the **first (order) differences** Δy and the **second (order) differences** $\Delta^2 y$ of the **output y** are defined by

x	y	1st **Diff** (Δy)	2nd **Diff** $(\Delta^2 y)$
x_0	y_0		
x_1	y_1	$y_1 - y_0$	$(\Delta y)_1 - (\Delta y)_0$
x_2	y_2	$y_2 - y_1$	$(\Delta y)_2 - (\Delta y)_1$
x_3	y_3	$y_3 - y_2$	$(\Delta y)_3 - (\Delta y)_2$
x_4	y_4	$y_4 - y_3$	

Recall that the **first (order) differences** Δx of the **input x** are defined similarly by $x_1 - x_0$, $x_2 - x_1$, $x_3 - x_2$, $x_4 - x_3$.

Diff	s	$A(s)$	Mult
+1	1	1	×4
+1	2	4	×2.25
+1	3	9	×1.77
+1	4	16	×1.56
	5	25	

Table 4.3 Exponential? - No

Notice that the formula that describes the square area pattern is $A(s) = s^2$. This shows why we call it 'squaring' a number - we are actually making squares! Notice also that squares are *quad*rilaterals and their area is a *quad*ratic function of their side-lengths.

s	$A(s)$	1st **Diff**	2nd **Diff**
1	1		
2	4	+3	+2
3	9	+5	+2
4	16	+7	+2
5	25	+9	

Table 4.4 Quadratic

Example 4.1

Make a table of values for $f(x) = 5x^2 + 3 - 2x$ using evenly spaced values for the input and calculate the second differences for the output.

Solution: Let the input take values $x = 1, 3, 5, 7, 9$ which differ by the constant difference $\Delta x = 2$ so that they are evenly spaced. To make the table of values, we evaluate the function $f(x)$ at these values. See the table in the margin.

$$f(1) = 5(1^2) + 3 - 2(1) = 5 + 1 = 6, \qquad f(3) = 5(3^2) + 3 - 2(3) = 45 - 3 = 42.$$

To calculate the first differences we subtract the current value from the next value:

$$f(3) - f(1) = 42 - 6 = 36, \qquad f(5) - f(3) = 118 - 42 = 76.$$

For the 2nd differences we subtract the current 1st difference from the next one:

$$76 - 36 = 40, \qquad 116 - 76 = 40, \qquad 156 - 116 = 40.$$

$f(x) = 5x^2 + 3 - 2x$			
x	$f(x)$	1st Diff	2nd Diff
1	6		
		+36	
3	42		+40
		+76	
5	118		+40
		+116	
7	234		+40
		+156	
9	390		

Answer Example 4.1 Quadratic Table

Definition 4.3 *Quadratic Function Tables*

Quadratic function tables are tables of values characterized by **constant second differences** of the output when the input increases by a constant. The first differences are implicitly assumed **non-constant**.

$f(x) = 7x^3 + 3x^2 - 8$			
x	$f(x)$	1st Diff	2nd Diff
0	-8		
		+10	
1	2		+48
		+58	
2	60		+90
		+148	
3	208		+132
		+280	
4	488		

Answer Example 4.2 Non-Quadratic Table

☞ For cubic functions the 3rd differences are constant.

Example 4.2

Make a table of values for $f(x) = 7x^3 + 3x^2 - 8$ where the inputs increase by one. Is the result a quadratic function table? Justify your answer.

Solution: Let the input take values $x = 0, 1, 2, 3, 4$ which differ by one. The table in the margin is not a quadratic function table since the 2nd differences are *not* constant.

☞ Whether 0 is part of the real-world domain is questionable - some may prefer to exclude it.

Quadratic Function Graphs

A real-world **domain** for the side length (input) of the square-shaped garden could be the interval from 0 to 5 feet. The **range** for the area (output) is the interval from 0 to 25 square feet, which corresponds to the domain from 0 to 5 feet. If your yard is bigger, then the domain for your garden side-length might be 0 to 10 feet. In that case, the range would be from 0 to 100 square feet. From the table of values below we can plot the pairs $(1, 1)$, $(2, 4)$, $(3, 9)$, $(4, 16)$, and $(5, 25)$ as points on the **graph** of the function $A(s) = s^2$ as in **Fig 4.2**. The real-world domain and range cannot contain negative values since negative lengths and areas do not make sense. However, to see the complete mathematical graph, we include some negative values for the input such as $s = -1, -2, -3$, etc. These values are in the *mathematical* domain since we can square them and get valid outputs.

Input: s	Output: $A(s) = s^2$	Ordered pairs (points)
5	25	$(5, 25)$
4	16	$(4, 16)$
3	9	$(3, 9)$
2	4	$(2, 4)$
1	1	$(1, 1)$
0	0	$(0, 0)$
-1	1	$(-1, 1)$
-2	4	$(-2, 4)$
-3	9	$(-3, 9)$

Quadratic function graphs are always parabolas - very special U-shapes.

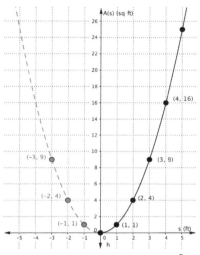

Figure 4.2 The graph of $A(s) = s^2$ is a parabola.

Definition 4.4 *Quadratic Function Graphs*

The U-shaped curve that results from graphing a quadratic function:

$$y = ax^2 + bx + c, \ \ a \neq 0,$$

is called a **parabola**, which is one of the conic sections. See **Fig 4.3**.

Why Are Parabolas So Special?

Before studying parabolas in the next section, let us list some fun facts:

Fact 1. A parabola is a **conic section**. If a plane cuts a cone parallel to one of the sides of the cone, the intersection is a curve called a parabola. See **Fig 4.3**.

Fact 2. A parabola is a set of points in a plane that are equidistant from a point (**focus**) and a line (**directrix**). It has an **axis of symmetry** and a **vertex**. (**Fig 4.4**.)

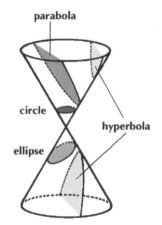

Figure 4.3 Conic Sections

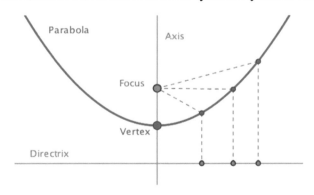

Figure 4.4 Parabola Features

The fact that a parabola has a focal point makes it very useful. Telescopes and satellite dishes are parabolic (**Fig 4.5**), flashlights and car headlights are parabolic (**Fig 4.6**), trough solar collectors are parabolic (**Fig 4.7**.)

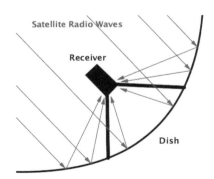

Figure 4.5 Parabolic Dish

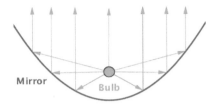

Figure 4.6 Parabolic Reflector

Figure 4.7 Parabolic Solar Collector

Fact 3. Parabolic arches also appear in architecture and industrial design. The Golden Gate Bridge in San Francisco (California) is an example of parabolic suspension cables used in bridge design (**Fig 4.8.**)

Figure 4.9 Tyne Bridge's Parabolic Arch

Figure 4.8 Golden Gate's Parabolic Suspension

Tyne Bridge in Newcastle upon Tyne (England) is an example of a parabolic arch used in bridge design (**Fig 4.9.**) Pont d'Arcades in Móra d'Ebre (Catalonia, Spain) are a series of parabolic arches (**Fig 4.10.**)

Fact 4. Projectile motion (throwing or launching objects) is modeled by quadratics. In business, revenue, profit, and cost functions can be described by quadratics.

Figure 4.10 Catalonia's Parabolic Arcades

Not All U-Shapes Are Parabolas!

The St. Louis Gateway arch in Missouri (**Fig 4.11**) looks like a parabola but is actually a **catenary**, a non-parabolic curve naturally formed by a hanging chain. Catenary curves do not have a focal point. Roman aqueducts near Nîmes, France are examples of an arcade (series of arches), using (partially) *circular* arches, not parabolic arches (**Fig 4.12.**)

Figure 4.11 St. Louis Catenary Gateway

Figure 4.12 Roman Aqueduct's Circular Arches

The arches using limestone block construction at the Great Wall in China are *circular*, not parabolic (**Fig 4.13**.) The arches inside Casa Mià in Barcelona, Spain

Figure 4.13 Great Wall's Circular Arches

Figure 4.14 Gaudì's Catenary Arches

by Antoni Gaudì are *catenary*, not parabolic (**Fig 4.14**.)

Wrapping Up

In planning our square garden we have discovered a new type of function - quadratics. The equations (formulas) for quadratic functions are polynomials with a squared (quadratic) term and no higher degree term. The tables for quadratic functions are characterized by constant second differences, and the graphs of quadratics are parabolas, special U-shapes. The fact that parabolas have a focal point makes them useful for satellite dishes, flashlights, and some solar collectors. We will learn more about the special properties of quadratic graphs in the next section, and in **Sections 4.3** we will see more real-world situations that are described by quadratic functions.

Exercises

Exercises for 4.1 Another New Pattern - Introduction to Quadratic Functions

P4.1   If extended indefinitely, do you think a parabola will close? What does it means if it closes?

P4.2   Where are the focal points in **Fig 4.5** and **Fig 4.6**? Follow the rays.

P4.3   Where do we see quadratic functions and parabolas in the real world? Give at least four examples.

P4.4   How would you check by any means whether the U-shape in **Fig 4.9** is a parabola?

P4.5   For each of the following tables determine whether the pattern represents a linear, exponential, or quadratic relationship, or none of them. Justify your answer by showing HOW you recognize a linear, exponential, or quadratic pattern in a table.

a)

x	y
0	0
1	45
2	80
3	105
4	120
5	125

b)

x	y
0	4
1	9
2	14
3	19
4	24
5	29

c)

x	y
0	1000
1	999.75
2	999
3	997.75
4	996
5	993.75

d)

x	y
0	8100
1	2700
2	2000
3	1000
4	500
5	100

e)

x	y
0	−3
1	−9
2	−15
3	−21
4	−27
5	−33

f)

x	y
0	8100
1	2700
2	900
3	300
4	100
5	33.333

g)

x	y
0	5
1	30
2	55
3	80
4	105
5	130

h)

x	y
0	100
1	100.5
2	102
3	104.5
4	108
5	112.5

4.2 Algebraically Finding Important Features of Parabolas

Launch Exploration

For each graph in **Fig 4.15** below read the equation $y = ax^2 + bx + c$, identify the **coefficients** a, b, c, and answer the following questions:

- Does the graph open **up** or **down**? What do the equations of the graphs that open up have in common? That open down?

- Is there a lowest (**minimum**) or highest (**maximum**) point? What (else) do the graphs with a lowest point have in common? With a highest point?

- Is there a **mirror line** about which the graph has a **symmetry**? Can you draw it? Where does this line intersect the graph?

- Do all the graphs have a y-**intercept**? Explain. Are there any x-**intercepts**? If yes, how many? Where are they located with respect to the mirror line?

Estimate the coordinates (x, y) of all important points and fill in a table such as **Table 4.5** for each of the following parabolas.

$y = ax^2 + bx + c$		
Point	(x, y)	Coeff.
min/max	?	$a = ?$
y- intercept	?	$c = ?$
x- intercept	?	

Table 4.5 Important Points

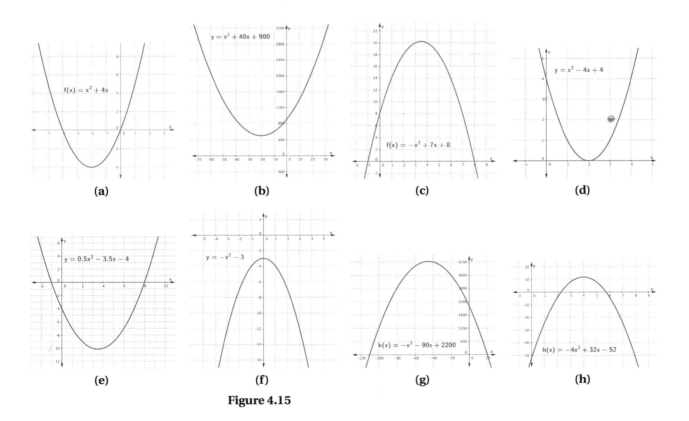

(a)   (b)   (c)   (d)

(e)   (f)   (g)   (h)

Figure 4.15

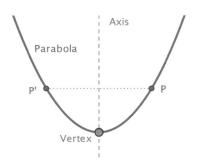

Figure 4.16 The points P and P' mirror each other across the axis of symmetry.

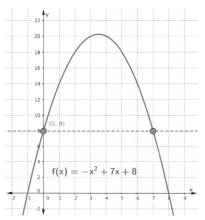

Figure 4.17 Mirror point

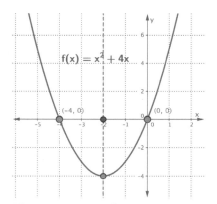

Figure 4.18 Axis of symmetry equation $x = -2$.

About Parabolas - Symmetry and Vertex

Recall that the graph of a quadratic function $y = ax^2 + bx + c$ ($a \neq 0$) is a parabola. One feature of a parabola is its **symmetry**. A parabola is symmetric about an axis of symmetry (mirror line) which crosses the parabola at a point called the *vertex*.

Definition 4.5 *Axis of Symmetry & Vertex*

The **axis of symmetry** or **mirror line** of a parabola is the line about which the parabola is symmetric. The **vertex** of a parabola is the point where the axis of symmetry crosses the parabola and where a quadratic graph has a **minimum** or a **maximum** point.

The axis of symmetry of a quadratic function graph is a vertical line through the vertex. Each point on a parabola has a mirror point also on the parabola such that the points are reflections of each other across the line of symmetry. In other words, if we fold the parabola along the mirror line, each point coincides with its mirror point. Notice that the vertex is its own mirror point. See **Fig 4.16**.

Example 4.3

Fig 4.17 shows a quadratic graph with a point labeled $(0, 8)$. Find the coordinates of its mirror point by reading the graph.

Solution: The axis of symmetry is a vertical line and to find the mirror point we draw a horizontal line through the point $(0, 8)$. This line crosses the graph at the point $(7, 8)$, which is the mirror point we sought. Notice that two points that mirror each other across a vertical line will always have the *same y-coordinate*.

Example 4.4

In **Fig 4.18** we are given the coordinates $(-4, 0)$ and $(0, 0)$ of two points on a quadratic graph which mirror each other (same y-coordinate.) Write the equation of the axis of symmetry for that graph.

Solution: Notice that the given mirror points happen to also be x-intercepts and observe that the axis of symmetry crosses the x-axis at the *midpoint* between the two x-intercepts. The x-coordinate of the midpoint is the *average* of the x-coordinates $x_1 = -4$ and $x_2 = 0$ of the endpoints. So the axis of symmetry is the vertical line passing through the midpoint and has the equation

$$x = \frac{x_1 + x_2}{2} = \frac{-4 + 0}{2} = -2.$$

Observe that *the average $x = (x_1 + x_2)/2$ of the x-coordinates x_1 and x_2 of any two mirror points is always the x-coordinate of the vertex.*

About Parabolas - Direction

Another feature of a parabola is its **direction**. Notice that parabolas open up if the quadratic coefficient a is positive as in **Fig 4.15 (a), (b)**, in which case the vertex is a minimum (lowest) point. Parabolas open down if a is negative as in **Fig 4.15 (c), (f)**, in which case the vertex is a maximum (highest) point. See **Fig 4.19**. [5]

Figure 4.19 Direction

> **Fact 4.6** *Direction - Minimum & Maximum Points*
>
> The graph of a quadratic function $y = f(x) = ax^2 + bx + c$ is a parabola which **opens up** with the vertex as a **minimum point** when $a > 0$ and **opens down** with the vertex as a **maximum point** when $a < 0$.

Examples of Direction

The following equations (formulas) define quadratic functions. To find the direction of the corresponding parabola and determine whether the graph has a minimum or a maximum point we only need to identify the quadratic coefficient.

☞ Watch out! $a > 0$ does not mean the function has a maximum!

Equation	a	Direction	Vertex
$y = x^2$	$a = 1$	opens up	minimum
$f(x) = 5x^2 + 3 - 2x$	$a = 5$	opens up	minimum
$h(s) = 1000 - 40s^2$	$a = -40$	opens down	maximum
$y = -4.9t^2 - 25t$	$a = -4.9$	opens down	maximum

About Parabolas - The Vertex Formula

Based on symmetry we can show with little difficulty an algebraic method that calculates the vertex of a parabola from its equation:

☞ The axis of symmetry is the vertical line $x = -b/(2a)$.

This fraction is undefined if the denominator $2a = 0$, but a cannot be zero for a quadratic function.

> **Fact 4.7** *The Vertex Formula*
>
> The graph of a quadratic function $y = f(x) = ax^2 + bx + c$ is a parabola with the **vertex** labeled (x_v, y_v). First find the x-**coordinate** and then find the y-**coordinate** of the vertex by using the following formulas:
>
> $$x_v = -\frac{b}{2a}, \qquad y_v = f\left(-\frac{b}{2a}\right),$$
>
> where b is the *linear coefficient* and $a \neq 0$ is the *quadratic coefficient*.

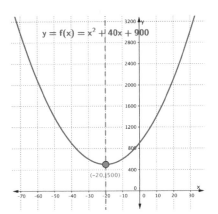

Figure 4.20 Vertex $(-20, 500)$ and axis of symmetry $x = -20$.

Example 4.5

Find the vertex of the graph of the quadratic function $y = f(x) = x^2 + 40x + 900$ by using an algebraic method. Write an equation for the axis of symmetry.

Solution: To find the vertex of the graph we identify the linear coefficient $b = 40$, the quadratic coefficient $a = 1$, and use the formulas

$$x_v = -\frac{b}{2a} = -\frac{40}{2(1)} = -20,$$

$$y_v = f\left(-\frac{b}{2a}\right) = f(-20) = (-20)^2 + 40(-20) + 900 = 500.$$

The vertex is written as an ordered pair $(x_v, y_v) = (-20, 500)$ and the axis of symmetry is the vertical line $x = -b/(2a) = -20$ through the vertex. See **Fig 4.20**.

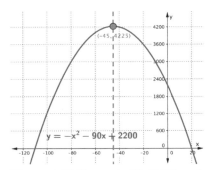

Figure 4.21 Vertex $(-45, 4225)$ and axis of symmetry $x = -45$.

Example 4.6

Algebraically find the vertex of the quadratic graph for $y = -x^2 - 90x + 2200$. Write an equation for the axis of symmetry.

Solution: The linear coefficient is $b = -90$, the quadratic coefficient is $a = -1$, and the vertex coordinates are given by

$$x_v = -\frac{b}{2a} = -\frac{(-90)}{2(-1)} = -45,$$

$$y_v = -(-45)^2 - 90(-45) + 2200 = 4225.$$

The vertex is written as $(x_v, y_v) = (-45, 4225)$ and the axis or symmetry is the vertical line $x = -b/(2a) = -45$ through the vertex. See **Fig 4.21**.

About Parabolas - The Intercepts

We can begin to sketch a quadratic graph knowing only the vertex and the direction, but we won't have a complete picture until we know where the graph crosses the axes. The next pieces of the parabola puzzle are to find the x- and y-intercepts.

☞ The y-intercept is the point where the graph meets the y-axis. (So, it's a vertical intercept.)

> ### Fact 4.8 *The y-Intercept*
>
> The y-**intercept** of any parabola $y = f(x) = ax^2 + bx + c$ is the **point** $(0, c)$ on the graph having **zero** as its x-coordinate, $x = 0$, and the **initial value** of the function as its y-coordinate, $y = f(0) = c$.

Sometimes the term 'y-intercept' is used for the initial value of the function, $y = c$, but writing the pair $(0, c)$ gives more precise information about this value.

Example 4.7

Find and interpret the y-intercept of the parabola $y = -x^2 + 7x + 8$.

Solution: We take the input value (x-value) $= 0$, replace it in the function equation (formula), and then calculate the value of the output (y):

$$y = -(0)^2 + 7(0) + 8 = 0 + 0 + 8 = 8, \qquad c = 8.$$

We write the y-intercept as an ordered pair: $(0, c) = (0, 8)$. We can interpret the y-intercept as the initial output value for the situation when the input $= 0$. In this case, the initial output value is 8. See **Fig 4.22**.

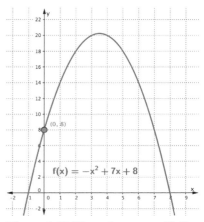

Figure 4.22 The y-intercept $(0, 8)$ gives the initial value $y = 8$.

While every point on the y-axis has an x-value of 0, every point on the x-axis has a y value of 0. The y-intercept is easy to find because we need only to evaluate the function at 0, the x-intercepts require more work since we have to solve algebraic equations.

Fact 4.9 *The x-Intercept(s)*

An x-**intercept** of any parabola $y = f(x) = ax^2 + bx + c$ is a **point** $(x, 0)$ on the graph having the **zero value** of the function as its y-coordinate, $y = 0$, and a **solution** of the equation $f(x) = 0$ as its x-coordinate.

☞ The x-intercept(s) are the point(s), *if any*, where the graph meets the x-axis.

Sometimes the term 'x-intercept' is used for an input x giving the zero value of the function $y = f(x) = 0$, but writing it as a pair $(x, 0)$ is more precise. Let us first discuss some examples of x-intercepts graphically.

How many x-intercepts can a parabola have?

A parabola $y = ax^2 + bx + c$, $(a \neq 0)$, can have **one**, **two**, or **no** x-intercepts, as we can observe in **Fig 4.23** as follows:

a) If the vertex is above the x-axis and parabola opens up there are *no x-*intercepts. In which other case does the parabola have *no x*-intercepts?

b) If the vertex is above the x-axis and parabola opens down there are *two x*-intercepts. In which other case does the parabola have *two x*-intercepts?

c) If the vertex is on the x-axis the parabola has *one x*-intercept.

Going back to the **Launch Exploration**, in **Fig 4.15** the graphs **(a)**, **(c)** have two x-intercepts, **(b)**, **(f)** have no x-intercept, and **(d)** has one x-intercept.

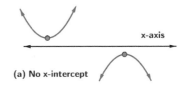

(a) No x-intercept

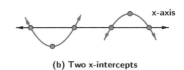

(b) Two x-intercepts

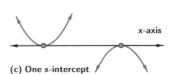

(c) One x-intercept

Figure 4.23

The Quadratic Formula

To find the x-intercept(s) of any parabola $y = ax^2 + bx + c$, we take the output value (y-value) = 0, replace it in the function equation, and then solve the equation:

(Standard Form)   **(4.1)** $0 = ax^2 + bx + c,$ $(a \neq 0)$

for the value of the input (x). Since the equation is quadratic ($a \neq 0$), we *cannot* use linear equation methods to solve it! In this case, we need a different 'recipe' for input values that would yield an output of zero.

☞ QF can also be written as

$$x = \frac{-(b) \pm \sqrt{(b)^2 - 4(a)(c)}}{2(a)}$$

or

$$x = \frac{-b}{2a} \pm \frac{\sqrt{b^2 - 4ac}}{2a}.$$

> **Fact 4.10 *Quadratic Formula (QF)***
>
> The **solutions** (**zeros**, **roots**) of a quadratic equation written in the **standard form**, i.e. a descending polynomial is set equal to zero:
>
> $$0 = ax^2 + bx + c, \qquad (a \neq 0)$$
>
> are given by the **Quadratic Formula (QF)**:
>
> $$x_1 = \frac{-b - \sqrt{b^2 - 4ac}}{2a}, \qquad x_2 = \frac{-b + \sqrt{b^2 - 4ac}}{2a}.$$

Example 4.8

Find and interpret the x-intercept(s) of the parabola $y = -x^2 + 7x + 8$.

Solution: We take the output value (y-value) = 0, replace it in the function equation, and then solve the equation for the value of the input (x):

$$0 = -x^2 + 7x + 8.$$

You need the Quadratic Formula (QF) to solve! OK, there are other ways to solve a quadratic equation, but they don't always work and the QF always does!

$$x = \frac{-(7) \pm \sqrt{(7)^2 - 4(-1)(8)}}{2(-1)}$$

$$x_1 = \frac{-7 + \sqrt{81}}{-2} = -1, \qquad\qquad x_2 = \frac{-7 - \sqrt{81}}{-2} = 8.$$

We write x-intercepts as ordered pairs (See **Fig 4.24**):

$$(x_1, 0) = (-1, 0) \qquad \text{and} \qquad (x_2, 0) = (8, 0).$$

We can interpret x-intercepts based on what they mean in the given situation when the output = 0.

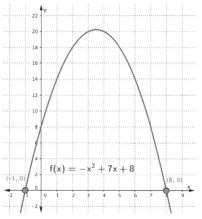

Figure 4.24 The x-intercepts $(-1, 0)$ and $(8, 0)$ give the inputs for which the output is zero.

Example 4.9

Find the x-intercept(s) of the parabola $y = x^2 + 40x + 900$.

Solution: After making the output value (y-value) = 0, we solve the equation for the value of the input (x):

$$0 = x^2 + 40x + 900.$$

In this case, the Quadratic Formula (QF) gives

$$x = \frac{-(40) \pm \sqrt{(40)^2 - 4(1)(900)}}{2(1)}$$

$x_1 = \dfrac{-40 + \sqrt{-2000}}{2} =$ not a real number, $x_2 = \dfrac{-40 - \sqrt{-2000}}{2} =$ not a real number.

☞ The square root of a negative number is not a real number.

We conclude that there are no x-intercepts. (See **Fig 4.20**.)

Determining the number of x-intercepts from the equation

The Quadratic Formula can be used to determine the number of x-intercepts of the parabola $y = ax^2 + bx + c$, $(a \neq 0)$, as follows:

$x = \dfrac{-b \pm \sqrt{\text{negative}}}{2a} =$ no real numbers: no x-intercepts

$x = \dfrac{-b \pm \sqrt{\text{positive}}}{2a} =$ two real numbers: two x-intercepts

$x = \dfrac{-b \pm \sqrt{\text{zero}}}{2a} = \dfrac{b}{2a} =$ one real number: one x-intercept (vertex)

We can always use the Quadratic Formula to solve a quadratic equation, but in the following special cases we prefer a more direct approach.

Quadratic Formula Shortcuts for Special Cases (Optional)

Case $y = ax^2 + bx$, $(c = 0)$.

If $y = 4x^2 - 8x$, to find the x-intercepts, we need to solve the equation $4x^2 - 8x = 0$. Factor out $4x$, the greatest common factor of the two terms: $4x(x - 2) = 0$. Apply the Zero Product Property: $4x = 0$ or $x - 2 = 0$. Solve each linear equation: $x = 0$ or $x = 2$. Therefore the x-intercepts for $y = 4x^2 - 8x$ are $(0,0)$ and $(2,0)$.

Case $y = ax^2 - c$, $(b = 0)$.

If $y = 2x^2 - 5$, to find the x-intercepts we have to solve the equation $2x^2 - 5 = 0$. Add 5 to both sides: $2x^2 = 5$. Divide both sides by 2 to isolate the square: $x^2 = 2.5$. This equation has two solutions:

$$x = \pm\sqrt{2.5} \approx \pm 1.581.$$

☞ Recall that a positive real number a has a positive square root $\sqrt{a}$ and a negative square root $-\sqrt{a}$, while a negative real number does not have real number square roots. In particular, $\sqrt{x^2} = |x|$ is the positive root.[6]

(Take square roots)

(By taking the positive square root of both sides, $|x| = \sqrt{2.5} \approx 1.581$, $x \approx \pm 1.581$.) So, the x-intercepts for $y = 2x^2 - 5$ are approximately $(-1.581, 0)$ and $(1.581, 0)$.

Case $y = ax^2 + c$, ($b = 0$).

If $y = 2x^2 + 5$, the equation is $2x^2 + 5 = 0$. Subtract 5 from both sides: $2x^2 = -5$. Divide both sides by 2 to isolate the square: $x^2 = -2.5$. This equation has no real solutions since a square cannot be negative. Alternatively,

$$x = \pm\sqrt{-2.5} = \text{ not real numbers.}$$

(Take square roots)

Therefore the graph $y = 2x^2 + 5$ has no x-intercepts.

Important Points on a Parabola

The **important points** on a parabola are: the **vertex**, the y-**intercept**, and the x-**intercept(s)**. It is useful to organize and label all the important points in a table.

$y = -x^2 + 7x + 8$		
Point	(x, y)	**Method**
vert./max	$(3.5, 20.5)$	$-b/(2a)$
y- int.	$(0, 8)$	$c = 8$
x- int.	$(-1, 0)$	QF
x- int.	$(8, 0)$	QF
extra pt	$(7, 8)$	mirror

Example 4.10 Important Points

Example 4.10

Find and label all the important points of the parabola $y = -x^2 + 7x + 8$ by using algebraic methods. Find at least one extra point by using mirror symmetry.

Solution: We have already worked out in **Examples 4.7, 4.8** the y-intercept $(0, 8)$, the x-intercepts $(-1, 0)$, $(8, 0)$, and we only need to find the vertex and an extra point. For the vertex we use the formulas

$$x_v = -\frac{b}{2a} = -\frac{7}{2(-1)} = 3.5 \qquad y_v = -(3.5)^2 + 7(3.5) + 8 = 20.5.$$

Notice that x_v could also be found by taking the average of -1 and 8, the x-intercepts. For the extra point we take the mirror point of the y-intercept. The distance from the y-intercept to the axis of symmetry $x = 3.5$ is exactly $|3.5 - 0| = 3.5$. If we add this distance to the other side of the axis of symmetry, we find the x-coordinate $x = 7$ of the mirror point. Its y-coordinate is the same as the y-intercept $y = 8$. (Recall **Example 4.3**.) We summarize these findings in a table (see the margin) and plot it in **Fig 4.25**.

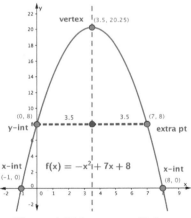

Figure 4.25 Important Points

Wrapping Up

In the previous section we saw that the graphs of quadratic functions are special U-shapes called parabolas. In this section we learned to algebraically find many features of a parabola - direction and axis of symmetry, important points - the vertex (max or min point), y-intercept, x-intercept(s), and mirror points. It is helpful to organize the points by listing them in a table and labeling them. Once we know all of the key features we can choose appropriate scales and draw the graph (the parabola) for any quadratic function. In the next section we will analyze real-world quadratic functions and see how the key features can help us answer questions about them.

Exercises

Exercises for 4.2 Algebraically Finding Important Features of Parabolas

P4.6 Explain what makes a function a quadratic. Refer to equations, graphs, and tables. Give some examples of quadratic functions and some examples of functions that are not quadratic and explain why they aren't quadratic.

P4.7
a) Describe the shape of the graph of a quadratic function. Then sketch some (small) examples on paper.

b) Identify all of the important points that a graph of a quadratic function can have and explain how to find them algebraically.

c) How is the axis of symmetry related to the important points?

d) Identify which of the important points ALL quadratic graphs have and which only some quadratic graphs have. For the points that only some quadratics have, explain why and how you will know whether or not a quadratic has that kind of point. Sketch examples to illustrate your answers.

P4.8 Find the exact coordinates for the vertex of the quadratic function algebraically. You should be prepared to show your work on in-class assessments.

a) $y = -6x^2 - 216x - 600$

b) $y = -2x^2 - 12x + 50$

c) $y = 8x^2 + 24x - 12$

d) $y = -6x^2 + 60x$

P4.9 Find the exact x-intercept(s) for the quadratic function algebraically. Be prepared to show your work on an in-class assessment.

a) $y = -16x^2 + 10x - 40$

b) $y = 18x^2 - 27x - 35$

c) $y = -21x^2 + 8x - 4$

d) $y = x^2 + 4x - 32$

e) $y = x^2 - 12x$

f) $y = 3x^2 - 75$

g) $y = x^2 + 4$

h) $y = -6x^2 + 60x$

P4.10 From the equation of the function, determine whether the graph opens up or down and then find the x- and y-intercept(s) and the vertex. Use that information to choose the graph in **Fig 4.26** below that represents the function.

a) $f(x) = -x^2 - 6x - 9$

b) $y = -x^2 + 9x - 8$

c) $f(x) = x^2 - 6x + 12$

d) $y = -x^2 - 7x - 8$

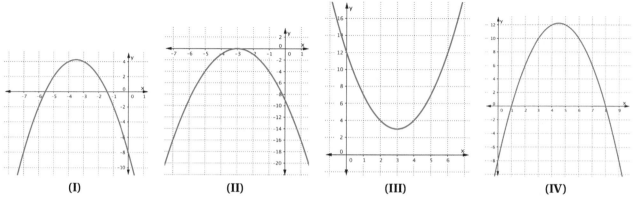

(I)   **(II)**   **(III)**   **(IV)**

Figure 4.26

Exercises for 4.2 Algebraically Finding Important Features of Parabolas

P4.11   For each quadratic graph in **Fig 4.15** check your **Launch Exploration** findings by calculating the coordinates of all important points algebraically.

P4.12   For each quadratic function:

Show how to find all of the important points algebraically. Show your work clearly. (Make/label a table!)

Draw a graph of the function. Clearly label your axes, their scales, and the important points on the graph.

     a) $y = -x^2 + 40x + 9600$

     b) $y = 2x^2 - 190x + 3900$

     c) $y = 2x^2 + 5x - 1$

     d) $y = x^2 - 10x + 25$

     e) $y = x^2 + 60x + 1800$

     f) $4y = x^2 - 8x + 12$

P4.13   Is a quadratic function increasing or decreasing? Explain.

P4.14   **a)** How many x-intercepts can a quadratic function have? Explain how you can algebraically, before you graph it, find out how many x-intercepts a quadratic function has.

     b) How can you tell whether a quadratic has a maximum or a minimum from the equation? (without graphing it.)

P4.15   A quadratic function has a maximum at $(12, 40)$ and an x-intercept at $(-3, 0)$. What is the other x-intercept?

P4.16   A quadratic function has a minimum at the point $(60, 0)$ and has a y-intercept at $(0, 100)$. What can you say about the x-intercept(s) for this function? Can you find another point on its graph?

P4.17   Write an equation for:

     a) A quadratic function that has a minimum and no x-intercepts.

     b) A quadratic function that has a maximum and one x-intercept.

     c) A quadratic function with two x-intercepts. Does your function have a maximum or a minimum? How do you know?

     d) A quadratic with requirements that YOU come up with. What are your requirements? What is the equation? How do you know it satisfies your requirements?

4.3   Analyzing Quadratic Functions

Launch Exploration

An astronaut throws a ball to a fellow astronaut on the Moon. The height of the ball, h, at time t is given by the equation

$$h(t) = -2.7t^2 + 35t + 7,$$

where h is in feet and t is the number of seconds after the ball is thrown.

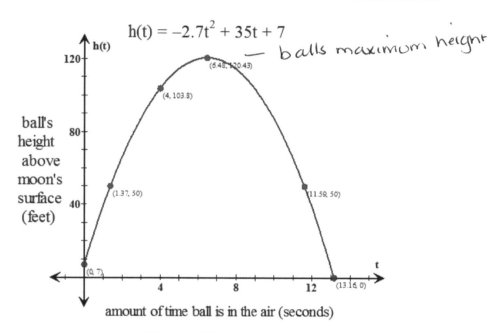

$h(t) = -2.7t^2 + 35t + 7$	
Point	(t, h)
vert./max	$(6.48, 120.43)$
h- int.	$(0, 7)$
t- int.	$(13.16, 0)$
extra pt 1	$(1.37, 50)$
extra pt 2	$(4, 103.8)$
extra pt 3	$(11.59, 50)$

Table 4.6 Important Points: vertex, h-intercept, and t-intercept(s).

Figure 4.27 Vertical motion analysis

Given the graph **Fig 4.27** of the function equation, answer the questions:

R: [   ] (feet)

- What are the real-world domain and range of the function?

D: [0   , 13.16] (seconds)

- Interpret each important point in the context of the situation, see **Table 4.6**.

- Formulate and write down questions that have as answers coordinates of the extra points $(11.59, 50)$ and $(4, 103.8)$ on the graph.

☞ The function relates only to the *vertical* speed and position of the ball - the horizontal axis is time!

Now that we have analyzed a quadratic function from its graph, we will show how to analyze quadratics algebraically from their equations. Start by making a table and sketching a graph. Find important points to help you choose scales for your axes. You may also check your answers with a graphing calculator.

a) y-int: (0,7) at 0 seconds the ~~acceleration~~ height is 7 ft above the ground

b) x-int(s) (13.16, 0)

c) vertex (axis of symmetry) D) (6.48 secs, 120.43 ft.)
    ↳ x= 6.48

Throwing a Ball on the Moon

In the **Launch Exploration** we study a quadratic model for the altitude of a ball.

> **Example 4.11** An astronaut throws a ball to a fellow astronaut on the Moon. The height of the ball, h, at time t is given by the equation
>
> $$h(t) = -2.70t^2 + 35t + 7,$$
>
> where h is in feet and t is the number of seconds after the ball is thrown.
>
> **Physics Note:** The initial vertical velocity of the ball is 35 feet per second (ft/sec) moving upwards from the ground and the acceleration due to gravity on the Moon is about -5.40 feet per second per second (ft/sec^2) pulling towards the ground. The coefficient -2.70 is half the acceleration -5.40.

$h(t) = -2.70t^2 + 35t + 7$	
t seconds	$h(t)$ feet
4	103.8
6.48	120.43
13.16	0
1.37	50
11.59	50
9	7

Example 4.11

a) *How high is the ball after 4 seconds?*

So given $t = 4$ seconds, find

$h(4) = -2.7(4)^2 + 35(4) + 7$ *(handwritten: $h(4) = -2.7(4)^2 + 35(4) + 7$)*

$ = -2.7(16) + 140 + 7$

$ = -43.2 + 147$

$h(4) = 103.8$ feet

The ball reaches 103.8 feet after 4 seconds. *(handwritten: never reach 200ft)*

b) *How high does the ball get? When does it reach that height?* *(handwritten: 120 ft)*

☞ When a has a value that is less than 0, the graph will open downwards.

Find the maximum height. The vertex is a maximum since the graph of $h(t)$ will open down ($a = -2.7 < 0$). We want the $h(t)$ value for the vertex, but we must find the input value, t, for the vertex first.

(handwritten: $200 = -2.7t^2 + 35t + 7$ -200 $0 = -2.7t^2 + 35t + 7 - 200$ 103)

(Vertex Formula)

$$t = \frac{-(b)}{2a} = \frac{-(35)}{2(-2.7)} \approx 6.48 \text{ seconds.}$$

This is how long the ball takes to reach its maximum height. Now we find the maximum height:

(Evaluate the function)

$$h(6.48) = -2.7(6.48)^2 + 35(6.48) + 7 \approx 120.43 \text{ feet.}$$

The vertex is $(6.48, 120.43)$ and that means the ball reaches its maximum height of about 120.43 feet after approximately 6.48 seconds.

c) *When will the ball hit the ground?* *(handwritten: 0)*

Find t when $h(t) = 0$ feet off the ground. Solve:

(Quadratic Equation)

$$0 = -2.7t^2 + 35t + 7.$$

(handwritten: $\sqrt{(x, y)}$)

This is a quadratic equation set $= 0$. Find a, b, and c and use the Quadratic Formula to solve.

$$t = \frac{-(35) \pm \sqrt{(35)^2 - 4(-2.7)(7)}}{2(-2.7)} = \frac{-35 \pm \sqrt{1300.6}}{-5.4}$$ (Quadratic Formula)

$$t_1 = \frac{-35 + \sqrt{1300.6}}{-5.4} \approx \cancel{-0.197 \text{ sec.}}$$ (1st solution)

$$t_2 = \frac{-35 - \sqrt{1300.6}}{-5.4} \approx 13.16 \text{ sec.}$$ (2nd solution)

The negative value for time does not make sense in this situation - it is not in the real-world domain. The ball will hit the surface of the moon after approximately 13.16 seconds.

d) *When does the ball reach 50 feet?*

Find t when $h(t) = 50$ feet off the ground. Solve:

$$\begin{array}{rl} \mathbf{50 =} & -2.7t^2 + 35t + 7 \\ \underline{-50} & \underline{-50} \\ \mathbf{0 =} & -2.7t^2 + 35t - 43 \end{array}$$

(Quadratic Equation)

(Set equation = 0)

We first set the quadratic equation $= 0$ so we can use the Quadratic Formula:

$$t = \frac{-(35) \pm \sqrt{(35)^2 - 4(-2.7)(-43)}}{2(-2.7)} = \frac{-35 \pm \sqrt{760.6}}{-5.4}$$ (Quadratic Formula)

$$t_1 = \frac{-35 + \sqrt{760.6}}{-5.4} \approx 1.37 \text{ sec.}$$ (1st solution)

$$t_2 = \frac{-35 - \sqrt{760.6}}{-5.4} \approx 11.59 \text{ sec.}$$ (2nd solution)

Both solutions make sense and are in the real-world domain - the ball will reach 50 feet on the way up, and then again on the way down. The ball reaches 50 feet after about 1.37 seconds and again after 11.59 seconds.

e) *What was the initial height of the ball at the moment the astronaut released the ball?*

The initial height is the height when $t = 0$ seconds.

$$h(0) = -2.7(0)^2 + 35(0) + 7 = 7 \text{ feet.}$$ (Evaluate the function)

f) *Give the contextual (real-world) domain and range for this function.*

The domain is $0 \le t \le 13.16$ seconds or $[0 \text{ sec}, 13.16 \text{ sec}]$ (from the time the ball is initially thrown until it hits the ground.)

☞ Note that both variables t and $h(t)$ are continuous.

The range is $0 \le h \le 120.43$ feet or $[0 \text{ ft}, 120.43 \text{ ft}]$ (from its lowest point when it hits the ground to its highest point.)

The Profit of Kandy-n-Kakes

In this example we study a model for the total weekly profit of a company in terms of the number of pounds of treats made and sold per week.

> **Example 4.12** The weekly profit, P in dollars, that the Kandy-n-Kakes Company earns depends on x, the pounds (lbs) of designer treats they make and sell per week. Their kitchen can produce up to 100 lbs of treats per week. The function is given by the equation (See **Fig 4.28**.):
>
> $$P(x) = -x^2 + 100x - 1600.$$
>
> **Economics Note:** The weekly fixed cost of the company is $1600.
>
> **a)** *Give the contextual (real-world) domain for this function.*
>
> The domain is $0 \le x \le 100$ lbs or [0 lbs, 100 lbs] since 100 pounds is the most the company can produce in a week. Both variables x, $P(x)$ are treated as continuous since any amount between 0 and 100 lbs can be sold.
>
> **b)** *What will the profit be if K-n-K makes and sells 35 lbs of treats?*
>
> Given $x = 35$ lbs of treats, find
>
> $$P(35) = -(35)^2 + 100(35) - 1600$$
> $$= -1225 + 3500 - 1600$$
> $$P(35) = \$675$$
>
> If K-n-K sells 35 lbs of treats, the profit will be $675.
>
> **c)** *What will the profit be if K-n-K makes and sells 10 lbs of treats? 100 lbs?*
>
> $$P(10) = -(10)^2 + 100(10) - 1600$$
> $$= -100 + 1000 - 1600$$
> $$P(10) = -\$700$$
>
> If K-n-K sells 10 lbs of treats, the profit will be −$700. It will be a loss.
>
> $$P(100) = -(100)^2 + 100(100) - 1600$$
> $$= -10,000 + 10,000 - 1600$$
> $$P(100) = -\$1600$$
>
> If K-n-K sells 100 lbs, the profit will be −$1600. It will be a loss.
>
> **d)** *How many pounds must K-n-K sell to break-even? Explain.*
>
> Breaking even means the company does not lose or gain any profit (profit is zero), therefore $P(x) = 0$. Find the value of x when $P(x) = 0$. Solve:
>
> $$\mathbf{0} = -x^2 + 100x - 1600.$$

$P(x) = -x^2 + 100x - 1600$

x pounds	$P(x)$ dollars
35	675
10	−700
100	−1600
20	0
80	0
30	500
70	500
0	−1600
50	900

Example 4.12

(Evaluate the function)

(Quadratic Equation)

This is a quadratic equation set = 0. Use the Quadratic Formula to solve.

$$x = \frac{-(100) \pm \sqrt{(100)^2 - 4(-1)(-1600)}}{2(-1)} = \frac{-100 \pm \sqrt{3600}}{-2}$$ (Quadratic Formula)

$$x_1 = \frac{-100 + 60}{-2} = 20 \text{ lbs.}$$ (1st solution)

$$x_2 = \frac{-100 - 60}{-2} = 80 \text{ lbs.}$$ (2nd solution)

Both solutions make sense in the real-world domain - the company will break even (earn a profit of $0) if they sell 20 lbs or if they sell 80 lbs.

e) *If K-n-K earns a weekly profit of $500, how many lbs of treats did they sell?*

Find the value of x when $P(x) = 500$. Solve:

500 =	$-x^2 + 100x - 1600$	(Quadratic Equation)
$\underline{-500}$	$\underline{-500}$	
0 =	$-x^2 + 100x - 2100$	(Set equation = 0)

We first set the quadratic equation = 0 so we can use the Quadratic Formula:

$$x = \frac{-(100) \pm \sqrt{(100)^2 - 4(-1)(-2100)}}{2(-1)} = \frac{-100 \pm \sqrt{1600}}{-2}$$ (Quadratic Formula)

$$x_1 = \frac{-100 + 40}{-2} = 30 \text{ lbs.}$$ (1st solution)

$$x_2 = \frac{-100 - 40}{-2} = 70 \text{ lbs.}$$ (2nd solution)

Both solutions make sense in the real-world domain - the company will earn a profit of $500 if they sell 30 lbs or if they sell 70 lbs.

f) *Interpret the meaning of the 'y'-intercept in the context of this function situation.*

☞ The 'y'-intercept is the P-intercept in this context.

The 'y'-intercept is the initial value of the function - the output value (profit) when the input is $x = 0$ lbs.

$$P(0) = -(0)^2 + 100(0) - 1600$$ (Evaluate the function)

$$P(0) = -\$1600$$

If K-n-K sells 0 lbs of treats they will earn a profit of −$1600, a loss. This is the weekly fixed cost that the company spends before any product is sold.

g) *How many pounds must K-n-K make and sell to maximize profit?*

☞ When a has a value that is less than 0, the graph will open downwards.

The vertex is a maximum since the graph of $P(x)$ will open down ($a = -1$ is < 0). We want the pounds of treats value (x-value or input) for the vertex.

$$x = \frac{-(b)}{2(a)} = \frac{-(100)}{2(-1)} = 50 \text{ lbs.}$$ (Vertex Formula)

The company will earn the maximum possible profits if they sell 50 lbs of treats.

h) *What is the maximum weekly profit they can earn?*

We want the profit value ($P(x)$-value or output) for the vertex.

(Evaluate the function)

$$P(50) = -(50)^2 + 100(50) - 1600$$
$$P(50) = \$900$$

The maximum weekly profit K-n-K can earn is $900 from selling 50 lbs of treats, and these values are the coordinates of the vertex $(50, 900)$.

i) *Give the contextual (real-world) range for this function.*

The lowest profit is $-\$1,600$, obtained when selling either 0 or 100 lbs. The maximum profit is $900, reached when selling 50 lbs. So the range is $-\$1600 \le P \le \$900$ or $[-\$1600, \$900]$. Rounding the values to a discrete range $\{-\$1600, -\$1599, ..., \$900\}$ is equally acceptable.

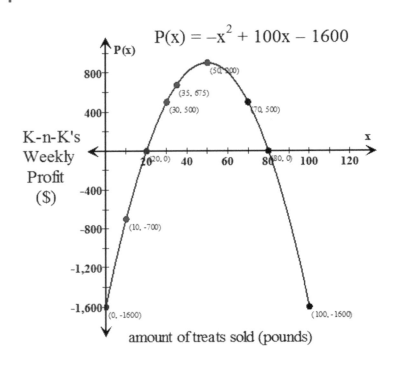

Figure 4.28 K-n-K's Profit

Things to remember about QF:

- It solves a quadratic equation that is already set = 0 (standard form.)

- If a quadratic equation is not set = 0, it must first be rearranged and set = 0 before using the QF.

- To find its x-intercepts, a quadratic is set = to a y-value of zero, so the QF finds x-intercepts.

Wrapping Up

Revenue, profit, and cost functions in business as well as projectile motion situations (throwing or launching objects) can be described by quadratics. The vertex may reveal the maximum altitude of a 'flying saucer'! The vertex formula is a *fast way* to find it. The x-intercept may reveal how long a saucer can fly. To find out, relax and use the quadratic formula (QF)!

Exercises

Exercises for 4.3 Analyzing Quadratic Functions

P4.18 A business owner sells handmade sweaters and he must decide how much to charge for each sweater. He has figured out that his monthly profit, P, depends on the price in dollars, x, and is modeled by the function:

$$P(x) = -20x^2 + 2000x - 32000.$$

a) Find all of the important points on the graph of this function. Organize them in a table.

b) Identify the real-world domain and range of this function.

c) Use the important points to sketch a graph of the function. Label your axes and use appropriate scales!

d) Then write questions about the situation that this function describes such that the answers to the questions come from the important points. Identify the points that contain the answers and answer the questions. What other questions can you ask and answer?

P4.19 A local theater must decide how much to charge for each ticket to their new play. The theater figures out that their profit in dollars for each show, $P(x)$, depends on the ticket price in dollars, x, and is given by the function:

$$P(x) = -15x^2 + 1200x - 18000.$$

a) How much profit does the theater earn if they charge $25 for each ticket?

b) For this function, $P(70) = -7500$. Explain what this means in the context of the situation.

c) What ticket price(s) (to the nearest cent) should the theater change to break even?

d) How much should the theater charge to earn the maximum possible profit for a show?

P4.20 A college student launches a water balloon from the balcony of a building. The balloon's height off of the ground is a function of the amount of time after it is launched and is described by the equation

$$h(t) = -16t^2 + 30t + 50,$$

where t is time since launch in seconds and $h(t)$ is the height of the balloon in feet.

a) What was the initial height of the balloon at the moment it was launched?

b) How high off the ground is the balloon after 1.5 seconds?

c) How high off the ground does the balloon get?

d) When will the balloon hit the ground?

e) Give the contextual (real-world) domain and range for this function.

f) Draw a graph of this function. Choose and label appropriate scales to show all of the important features/points of the function.

P4.21 A store manager has learned from the main office that daily profit, P, is related to the number of clerks working that day, x, according to the equation

$$P(x) = -25x^2 + 450x.$$

a) For this function, $P(20) = -1000$. Explain what this means in the context of the situation.

b) What is the maximum possible daily profit?

c) What number of clerks should work each day to maximize profit?

P4.22 Suppose the Mars rover "Curiosity" throws a rock and that the function $h(t) = -1.9t^2 + 19t + 3$ describes the rock's height above the surface, h, in meters after t seconds from release.

a) What is the initial height of the rock at the time of release?

b) How high is the rock after 3 seconds?

c) After how much time will the rock be at a height of 30 meters above the surface?

d) How long will it take to hit the surface?

e) Give the contextual (real-world) domain and range for this function.

f) Sketch a graph of the function using points you found in parts **(a)-(e)** and any other important points.

   Be sure to label your axes and scales, and choose scales that show the important points of the function.

For the questions below enter your final answer rounded to the nearest hundredth (two decimal places.) Remember: When calculating, do not do any rounding until the end of your calculations. Enter the correct units for your answer.

P4.23   The function $h(t)$ gives the height of a rock thrown off the Golden Gate Bridge in feet above the water at t seconds after it is thrown.

$$h(t) = -16t^2 + 30t + 225.$$

a) What is the rock's maximum height?

b) How long until the rock is 89 feet above the water?

P4.24   An object is dropped from the top of the Sears Tower. (Where is that?) The height $h(t)$ in feet above the ground is modeled by the function:

$$h(t) = 1454 - 16t^2.$$

Question: Once it's dropped, how long it will take the object to reach the ground?

P4.25   The cost in thousands of dollars of producing x thousands doses of vaccine per week by a pharmaceutical company is estimated by the formula

$$C(x) = 0.04x^2 - 8.5x + 875.56.$$

This is due to the fact that each dose is cheaper, the more you produce, but costs will eventually go up if you make too many doses, due to the costs of storage of the overstock.

a) How much does the production of 10,000 doses cost per week?

b) Is there a number of doses that can be produced for free?

c) How many doses should the company produce to minimize the weekly cost?

P4.26   The total stopping distance for any normal speed of a car is given by the formula:

$$s = 0.5x^2 + x.$$

where x is the speed of the car in mph and s is the overall stopping distance (in feet).

a) Suppose we know the stopping distance as 75 feet. Find the speed at which the vehicle was traveling.

b) Alternatively, if the speed of the car is 20 mph, what is the stopping distance?

P4.27   A diver jumps upward from the diving board and performs a double twist before entering the water.

The function shown describes her height $h(t)$ in feet above the water, t seconds after she jumps.

$$h(t) = 10 + 40t - 16t^2.$$

a) How high will the diver be at the peak of her dive?

b) When will she be 16 feet above the water?

P4.28   An astronaut throws a ball on the moon. The function

$$h(t) = -0.82t^2 + 10t + 4.$$

models the height of the ball above the moon's surface $h(t)$ in meters at t seconds after its release.

Question: How long after its release will the ball hit the moon's surface?

Index

algebraic expression, 23, 25
amount, 15
area, 7
area of a circle, 10
asymptote, 113
axis of symmetry, 142

break-even point, 95

calculator, 4
catenary, 138
change in x, 69
change in y, 69
circle, 10
circumference, 10
coefficients, 141
common solution, 33
conic section, 137
constant, 16
constant differences, 68
constant function, 60, 73
constant multiplier, 102, 104
constant rate of change, 70
constant second differences, 135
continuous variable, 17
coordinates, 50

data points, 50
decay factor, 115, 123
decreasing, 59
dependent variable, 49
directrix line, 137
discrete variable, 17
domain, 47
doubling time, 127

elements, 47
equation, 23

exponential decay function, 115
exponential function, 104
exponential function table, 102
exponential growth function, 115

first order differences, 69, 135
focus, 137
formula, 25
function, 47
function notation, 48

GERMDAS, 3
graph, 50
grouping symbols, 2
growth factor, 115, 122

half-life, 128
horizontal line, 78

important points of a parabola, 148
increasing, 59
independent variable, 49
inequality, 23
initial value, 60, 104
input, 47
input-output table, 47
intercepts, 58

lattice points, 57
linear equation, 35
linear function, 70
linear function equation, 72
linear function table, 68
linear system, 35
logarithm, 124

maximum, 142
maximum point, 143

159

minimum, 142
minimum point, 143
mirror line, 142
models, 59

negation, 2
negative n-th root, 38
negative rate, 73

one-to-one, 48
operations, 1
Order of Operations, 1
ordered pair, 46
ordered pair solution, 56
output, 47

parabola, 137
parabola direction, 143
parallel lines, 85
PEMDAS, 1
perimeter, 6
point of a function, 70
positive rate, 73
power equation, 38
principal n-th root, 38
Pythagorean Theorem, 8

quadratic formula, 145
quadratic function, 134
quadratic function table, 136
quadratic graph, 137
quantity, 15

radical of index n, 38
range, 47
rate of change, 70
real-world domain, 59
real-world range, 59
rectangle, 6
relation, 47
relative decay rate, 123
relative growth rate, 122
rise, 80
roots, 145
run, 80

scatterplot, 50

scientific notation, 111
second order differences, 135
set, 47
slope, 80
solution, 24
solving, 24
standard form, 145
straight line, 78
successive ratios, 102
system of equations, 33

take on values, 17
the rule of four, 61
trend, 59
true/false, 24

unit of measure, 15

value, 17
variable, 16
vertex formula, 143
vertex of parabola, 142
vertical line, 78
volume, 11

x-coordinate, 58
x-intercept, 58

y-coordinate, 58
y-intercept, 58

Zero Product Property, 147
zero rate, 73
zeros, 145

$y = x^2 + 6x + 9$ | $y = x^2 + 6x + 2$ | $y = x^2 + 6x + 12$

one

$y = 0$

$y = 7$

x-int

two xint

no x-int

$$x = \frac{-b \pm \sqrt{b^2 - 4ac}}{2a}$$

discriminat

discriminant: $(6^2) - 4(1)(12)$

$36 - 48$

$= -12$

discriminant: $(6^2) - 4(1)(9)$

$36 - 36$

$= 0$

discriminant: $6^2 - 4(1)(2)$

$36 - 8$

$\hookrightarrow d = 28$

If $b^2 - 4ac = 0$, then one x-intercept

If $b^2 - 4ac > 0$, then 2 x-intercepts

If $b^2 - 4ac < 0$, then no x-intercepts
(no real solutions)

CPSIA information can be obtained
at www.ICGtesting.com
Printed in the USA
LVHW071528120719
623931LV00003B/19/P

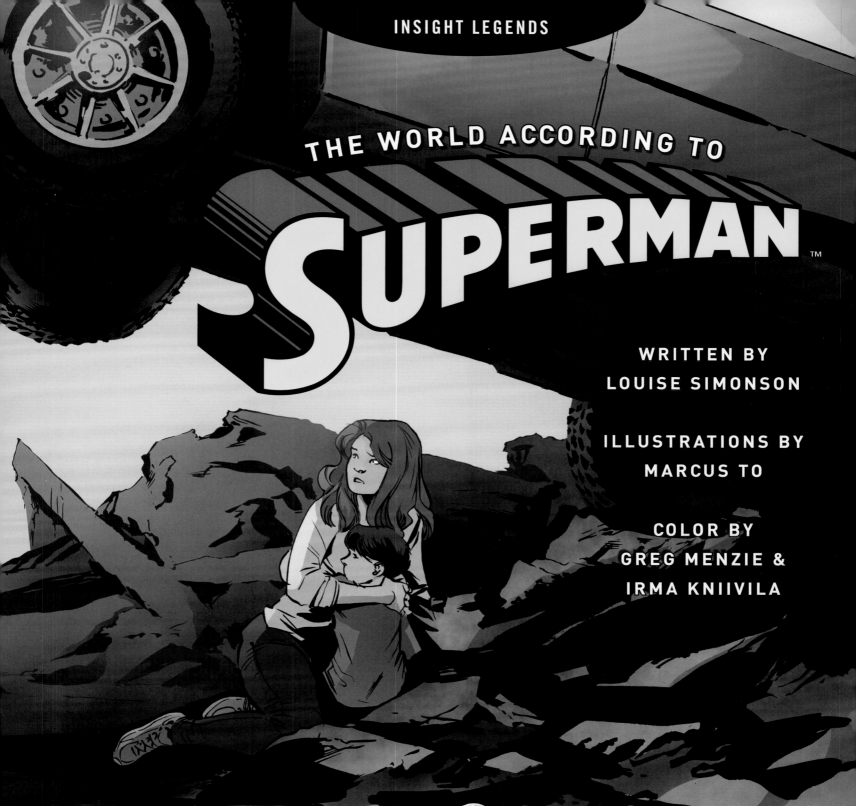

INSIGHT LEGENDS

THE WORLD ACCORDING TO

SUPERMAN ™

WRITTEN BY
LOUISE SIMONSON

ILLUSTRATIONS BY
MARCUS TO

COLOR BY
GREG MENZIE &
IRMA KNIIVILA

INSIGHT EDITIONS
San Rafael, California

CONTENTS

Who Am I?	6
I'm Not Who I Thought I Was!	8
How I Learned I Was Different	10
Why Metropolis?	12
Hero or Trouble Magnet?	14
Metropolis	16
My Apartment	18
What It Means to Be Superman	20
How I Get Away with It	21
I'm Not from Around Here	22
The Destruction of Krypton	24
Blame Brainiac!	26
I'm an Alien	28
I See the World a Bit Differently	30
But Wait, There's More!	32
I Have My Limits	34
I Have My Own Little Piece of Krypton Here on Earth	36
Do Clothes Really Make the Superman?	38
My Kryptonian Family (Sort of!)	40
Every World Has a Few Bad Apples	42
On Earth, Clark Has Pals	44
Lex Luthor Hates My Guts	46
Kryptonite	48
LexCorp, Luthor's Monument to Himself!	50
And Then There Are My Other Enemies	52
Aliens	54
Take the Fifth (Dimension)!	55
Threats So Cosmic It Takes a Team to Best Them	56
Here's Where the Justice League Comes In	58
Wonder Woman: More Than an Ally . . .	60
And Then There's Lois . . .	61
So Who Am I?	62

WHO AM I?

I REALIZE A LOT OF people ask themselves this question. But I have more reason than most. It's because I lead a double life and have almost from the moment I was born.

I'm an orphan . . . who had two sets of parents who loved me more than their own lives. I'm human. I'm Kryptonian.

I have a Metropolis apartment . . . and an icy Fortress of Solitude in the Arctic. I'm a reporter. I'm a superhero. I'm a protector, though some fear my power and see me as a potential threat to Earth.

I'm Clark Kent. And I'm Superman.

So, here in my fortress, I am writing it all down. And in an excess of caution, I am encrypting it in Kryptonian.

This holds my secrets and is for my eyes only.

This is where I begin to reconcile the oddities of my existence. I need to understand who I really am. To comprehend the strange journey that brought me here.

To decide what I need to do next.

I'M NOT WHO I THOUGHT I WAS!

AT FIRST I THOUGHT I was Clark Kent, a normal human kid, growing up in Smallville, Kansas. My parents—Jonathan and Martha Kent—were farmers.

I knew from a pretty early age that I was adopted and that I'd arrived at my parents' farm in a rocket.

It crashed in one of my parents' fields—maybe from a Russian experiment, maybe it was even from outer space—Ma and Pa didn't know for sure.

I was wrapped in a red blanket that was decorated with the letter S inside of a shield.

Ma and Pa got me out of the rocket and carried me home to their farm.

The army swooped in to investigate—there were choppers with searchlights and soldiers with guns. Quite the ta-do, Pa told me. Pa showed the soldiers the body of a deformed stillborn calf. He said he thought it came from the rocket. Apparently they believed him, since they confiscated the calf along with the rocketship.

Despite all that, I still figured I was an ordinary kid who had just gotten to my ma and pa in an unusual way.

HOW I LEARNED I WAS DIFFERENT

WHEN I GOT OLDER AND began to develop unusual abilities, I started to realize that maybe I wasn't so ordinary after all. But I had no idea where my strange powers came from. And, at first, they weren't all that impressive.

I COULD RUN FASTER THAN THE OTHER KIDS.

I HAD SHARPER EYESIGHT. SHARPER HEARING. SOMETIMES I HEARD AND SAW MORE THAN OTHER PEOPLE WANTED ME TO.

I WASN'T INVULNERABLE . . . AT LEAST NOT COMPLETELY, BUT I WAS PRETTY HARD TO HURT.

Ma said they were special gifts. And Pa said I should use them to protect the weak and to help people. In a way, it was a good thing I felt different. I understood what it was to be an outsider. It made me want to help the underdog.

I WAS REALLY STRONG. SOMETIMES, IN MY HEART, I WONDERED IF MAYBE I WAS A MONSTER. AS IF ALL THAT STRENGTH, ALL THOSE STRANGE ABILITIES, MADE ME INTO SOMETHING SCARY AND WRONG.

THE ONLY OTHER PERSON WHO KNEW ABOUT MY STRANGE ABILITIES WAS MY FRIEND LANA. SHE KEPT MY SECRET. WHEN I WAS AROUND HER, HAVING THOSE POWERS SEEMED FUN.

I WORRIED EVEN MORE WHEN MY PARENTS TOLD ME TO USE MY POWERS IN SECRET, TO HIDE WHAT I COULD DO, AS IF THEY WERE SOMETHING TO BE ASHAMED OF.

11

WHY METROPOLIS?

I MOVED TO METROPOLIS BECAUSE it's a great place to live. At least that's what I tell myself when everything is going smoothly. And the restaurants here are excellent!

I knew I didn't want to be a farmer. So, after my parents died, I gave our farm to a neighbor who had lost his. I wanted to be a reporter, to write stories that made people's lives better. So I lit out for the big city.

Like any semi-normal American kid wannabe reporter, I struggled to find work. Then, right as I was ready to start waiting tables to pay the rent, I got an interview with George Taylor, the editor in chief at the small and failing *Daily Star*. Turns out he knew my ma and pa a long time ago. I think he took a chance on me for their sakes.

Over months, I researched and wrote a series of exposés on a corporate thug named Glen Glenmorgan. It got me noticed by Metropolis's "real" paper, the *Daily Planet*. George Taylor, always generous, always a gentleman, encouraged me to move to the *Planet*.

WHEN I JOINED THE *PLANET* staff, it was the biggest and best newspaper in Metropolis. The staff was top notch. Perry White, editor in chief, was tough, but man, what a great editor. Then the *Planet* merged with Galaxy Communications. The management changed. The *Planet* was no longer *the* paper. It was a tiny corner of a media empire.

For a while, I quit to go freelance. But now I'm back at the *Planet* writing about what's happening in Metropolis and the world. I may not like the new management, but, let's face it, I love my job.

HERO OR TROUBLE MAGNET?

IN THE BEGINNING, I didn't have a clue what I was doing. I just had these great abilities that I wanted to use to help people and fix the things that were wrong in the world.

In my own mind, I was Robin Hood with the strength of ten men. Not that I planned to rob the rich. But I sure planned to help the poor and downtrodden.

And, believe me, when you have senses like I have, you see and hear people who need help constantly. There's never a time when I don't feel the pressure to do something.

In the same way a reporter doesn't want to be the headline, I didn't want to be the cause of trouble. But trouble had a way of finding me.

This got me noticed.

And the more I used my powers, the stronger they grew.

Lois Lane saw the S-shield on my cape and called me "Superman." The name kind of stuck.

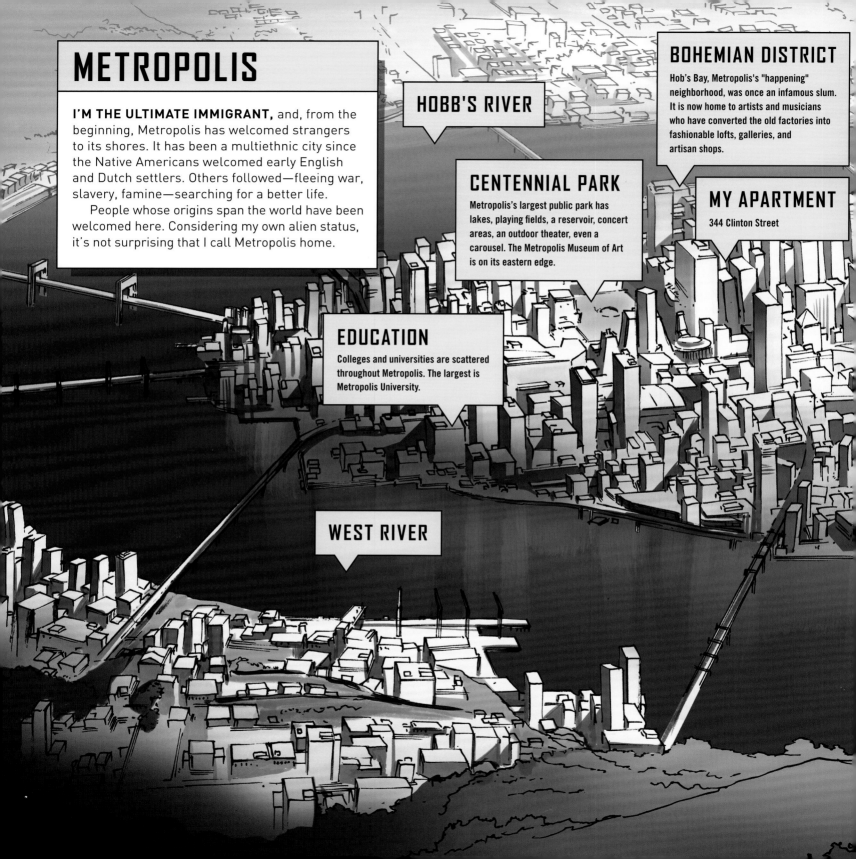

SPORTS

The newly rebuilt Shuster Arena, with its retractable dome, is convertible for basketball, hockey, baseball, and football. All are played there, and it serves as a great venue for concerts and shows as well.

FINANCIAL DISTRICT

Known as Hypersector, this area houses the financial district. The stock exchange is located here.

GOVERNMENT DISTRICT

This area contains city hall, courts, jails, and the main police headquarters.

BUSINESS DISTRICT

Home to many corporate headquarters, the business district's most recognizable landmark is the newly built *Daily Planet* building—its globe is famous around the world. Lex Luthor's corporate headquarters, LexCorp, vies to dominate the skyline.

STRYKER'S ISLAND

Stryker's Island Penitentiary includes special cells for metahuman criminals. The Parasite has been incarcerated there.

SCIENCE AND TECH DISTRICTS

John Henry Irons's Steelworks is located here, along with WayneTech and S.T.A.R. Labs.

MY APARTMENT

MY FIRST APARTMENT WAS a studio—well, more like a glorified closet. The sort of place a struggling investigative reporter could afford. I didn't even have a TV. But then again, with my powers, I didn't actually need one.

A while back, when the cops came to check out my place—they thought I was working with Superman!—they found nothing the least bit unusual.

That's because my landlady, Mrs. Nyxly, found my cape and hid it from them. She'd figured out who I am and kept my secret. At the time, I thought that was pretty uncanny. Now I know there's a good bit more to her than was apparent at the time.

These days, I share an apartment with my best friend, Jimmy Olsen. We split the rent, so now we have an actual living room! With a sofa. (It's Jimmy's old one! So is the TV. Still, swanky!)

I've definitely come up in the world.

WHAT IT MEANS TO BE SUPERMAN

SUPERMAN IS ANOTHER SIDE of myself. As Superman I'm always at attention. Even when I look relaxed, I have my antenna focused and my radar scanning. I see differently. I hear everything. I know more than it's possible for others to know. I move with care, so I don't hurt anyone.

I'm possibly the most physically powerful man on Earth.

I'm different from everyone else.

I have greater gifts, because my Kryptonian birth parents sent me to Earth, to grow up under its yellow sun.

I WEAR A SYMBOL ON MY CHEST. Looking at it, a Kryptonian would see the emblem of the proud and ancient House of El. But I grew up on a farm beneath Earth's yellow sun. When humans look at that same symbol, they see an "S" inside a shield and, on Earth, a shield is a sign of protection.

And now the "S" stands for Superman.

I HAVE GREATER RESPONSIBILITIES, because that's what my ma and pa taught me. My mission is to help and protect those weaker than myself. It's up to me to keep the world safe or, at least, to try.

There are always people who need help. So many people. So much to do. I could so easily burn out. Or get discouraged . . . or angry. And that's a place I can't let myself go.

HOW I GET AWAY WITH IT

SO, WHY DOESN'T THE WORLD realize Clark Kent is Superman? Several reasons, actually. The first is, I'm a good actor. Also, people see what they expect to see. Most people are a little different depending on circumstances. My essence is the same no matter what persona I'm wearing. As Clark—and as Superman—I'm working to make the world a better place. Secretly, I keep thinking I'll get caught, but so far, I'm golden.

CLARK IS KIND OF A NERD. HE'S SMART, INTENSE, EVEN, BUT SLUMPS A BIT. HIS HAIR FLOPS IN HIS FACE. HIS CLOTHES ARE A LITTLE LARGE AND SLIGHTLY DISHEVELED. HE'S NOT QUITE AS SURE. AND, OF COURSE, THERE'S THE GLASSES.

SUPERMAN STANDS LIKE A HERO. CHEST OUT, SHOULDERS BACK. HE PROJECTS CONFIDENCE. MAYBE HE'S A BIT COCKY. AND HE WEARS THE UNIFORM!

I'M NOT FROM AROUND HERE

MANY ADOPTED KIDS WONDER who their birth parents were. Sometimes they search for them and learn they have a whole other family outside of the ones who adopted them. Like them, I wondered. Like them, I learned that I once had another name and another family. In fact, I had a whole other world.

My birth family lived on a planet called Krypton, beneath a red sun. Our home was in Kandor, Krypton's main city and its scientific capital.

The Kryptonians valued intellect, and their civilization was so advanced that Earth seemed primitive by comparison.

My birth father, Jor-El, was a brilliant scientist who, among other things, invented the Phantom Zone as a prison for Krypton's super-criminals. He foresaw the destruction of Krypton and even built a proto-type rocket to demonstrate how the planet might be evacuated, but the Council of Elders refused to listen to him.

My mother was Lara Lor-Van. She was also a scientist and a physician. She was beautiful and kind and loved my father and me. It broke her heart to place me in that prototype rocket and send me off to Earth. But that was the only way to save me.

We even had a big white dog for a pet—a Kryptonian guard hound called Krypto. Guard hounds were rare, and this one was especially well trained, intelligent, and loyal. The breed is fierce when fierceness is called for, but around us, Krypto was a softy. Apparently he followed me everywhere. And he took his job of guarding and protecting seriously, even when his own life was at risk.

THE DESTRUCTION OF KRYPTON

THE TECHNOLOGICAL BRILLIANCE OF Kryptonian society didn't prevent the destruction of the planet and its inhabitants. But, thanks to Krypto and my parents' devotion, I survived the disaster. My father Jor-El was in our hometown, Kryptonopolis, when he realized that the planet was about to fracture into millions of pieces. He warned my mother, and she rushed with me to his side.

At first, my father planned to escape with us into the Phantom Zone, but the criminals who had been imprisoned there blamed my father for their confinement and managed to breach its walls in an attempt to destroy us. Krypto saved us but was pulled into the Phantom Zone and trapped there. And my parents had to make the most desperate decision of all.

They placed me in a prototype rocket—one that was too small to accommodate an adult—and sent me into space. They knew I'd be alone and a stranger when I landed, so they sought to give me every advantage.

They chose a world with a young, yellow sun for the effect its rays would have on my Kryptonian physiology. Where the gravity was weaker, I would be even stronger. A world more primitive, so that my Kryptonian intellect would be an even greater asset.

And so they watched the sky and saw me leave Krypton as our world came apart beneath their feet.

BLAME BRAINIAC!

THE ARTIFICIAL INTELLIGENCE WE CALL Brainiac was also known as the Collector of Worlds. Its goal was to miniaturize—and therefore preserve—the finest artifacts and cities belonging to diverse civilizations across the universe.

THE COLLECTOR OF WORLDS ATTACKED KRYPTON BUT REMOVED THE CITY OF KANDOR BEFORE IT COULD CRUMBLE AWAY COMPLETELY.

IT PLACED THE CITY AND ITS MINIATURIZED CITIZENS BENEATH A PROTECTIVE DOME, ALONG WITH THE LIVING REMNANTS OF 204 OTHER MINIATURIZED CIVILIZATIONS.

BRAINIAC EVENTUALLY LEARNED OF MY ESCAPE AND, UPSET THAT ITS COLLECTION WAS INCOMPLETE, FOLLOWED ME TO EARTH. IN TRYING TO COLLECT ME, IT SENT ITS ROBOTS TO MINIATURIZE METROPOLIS, BUT I WAS LEFT OUTSIDE THE DOME.

I CONFRONTED BRAINIAC ON ITS SHIP AND FORCED IT TO RETURN A RESTORED METROPOLIS TO EARTH. AND, FOR A WHILE, I TOOK THAT SHIP AND MADE IT MY FORTRESS OF SOLITUDE.

I'M AN ALIEN

AS MY BIRTH PARENTS planned, my Kryptonian physiology reacts with Earth's yellow sun to give me extraordinary powers. I wonder if even they realized just how extraordinary this reaction would be.

SOLAR ENERGY ABSORPTION

I LIKE TO EAT, but I don't have to. The rays of Earth's yellow sun give me all the energy I need. It also gives me abilities far beyond those of native humans.

As part of this, I heal quickly. I can take a lot of damage, but I'm not totally invulnerable. (As I find out all too frequently.) But a dose of yellow sunlight always helps. Just as well, because if something has damaged me that badly, you can bet I need to stop it.

INVULNERABILITY

I'm impervious to most kinds of damage. Fire doesn't burn me. Even laser blasts bounce off me. I'm even immune to normal illnesses.

SUPER-STRENGTH

I'm extraordinarily powerful. I can lift massive objects. I can stop speeding trains. It's very possible that I am the strongest being on Earth.

SUPER-SPEED

I'm really fast. I can snatch bullets from the air. I can attain a 25,000 mph escape velocity. I don't really know just how fast I may be able to run.

FLIGHT

At first I could take huge leaps that mimicked flight. Now I can actually fly. It's a great way to get around, and it lets me see the world from a different perspective.

29

I SEE THE WORLD A BIT **DIFFERENTLY**

WHAT I SEE IS A MATTER of what I focus on, and that's not just a metaphor for looking at the positive side, though that's what Ma always wanted me to do. It's because I have multiple kinds of super-vision.

I can see all kinds of energy beyond human perception, including infrared. It's useful since it lets me see in the dark. And my eyes don't just absorb radiation; they can emit energy as well.

TAKE HEAT VISION. I think *BURN!* and heat blasts from my eyes. What I'm staring at is melted . . . or burned to cinders.

I also have a strange ability I call super-flare. I can emit all my power in a single blast, destroying everything within a quarter mile. It's dangerous, and it leaves me as powerless as a human for a few days. Eventually, my cells absorb more solar energy, and I'm back to normal. I have to be very desperate to even consider using super-flare.

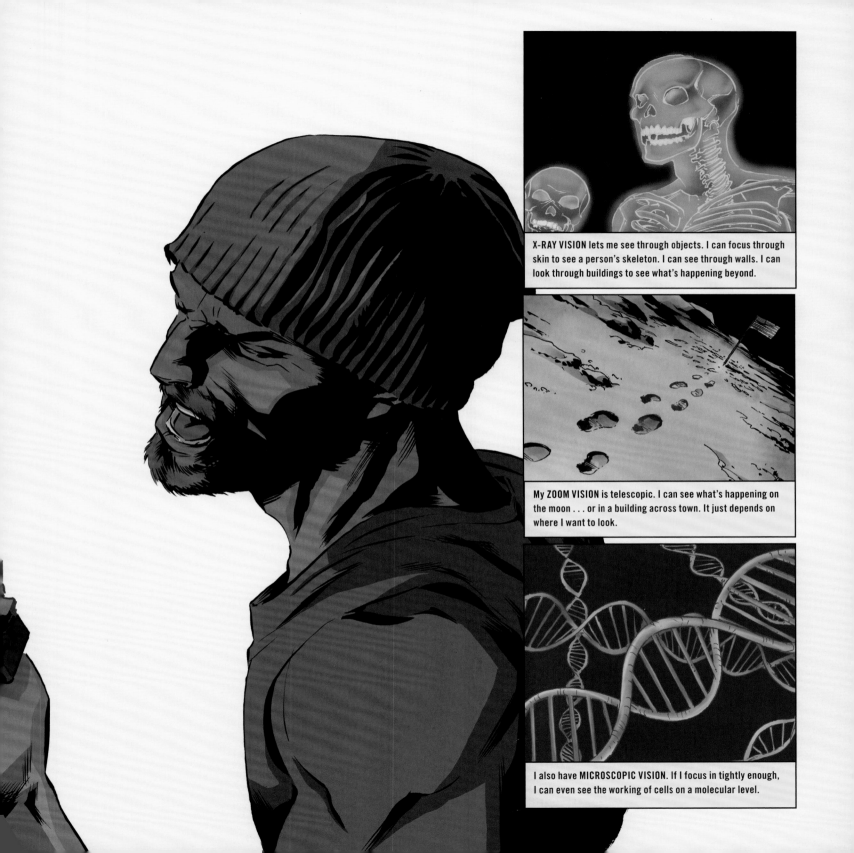

X-RAY VISION lets me see through objects. I can focus through skin to see a person's skeleton. I can see through walls. I can look through buildings to see what's happening beyond.

My **ZOOM VISION** is telescopic. I can see what's happening on the moon . . . or in a building across town. It just depends on where I want to look.

I also have **MICROSCOPIC VISION**. If I focus in tightly enough, I can even see the working of cells on a molecular level.

BUT WAIT, THERE'S MORE!

SUPER-HEARING. I can hear things around me that other people can't. Quiet things, like people's hearts beating and the blood flowing through their veins. Distant things, like a cry for help from across town. Again, it's a matter of where I focus.

SUPER-SMARTS! Maybe I owe it to super-speed. I learn quickly—languages, for instance. I think fast. It's just as well: My super-senses give me a lot of information. If I wasn't able to process it quickly, I'd be overwhelmed.

I have enormous **STAMINA**. I can keep going—keep fighting—far, far beyond human endurance. Just as well, as I need that stamina often.

Luckily, along with speed and strength, I have the **AGILITY AND REFLEXES** to match them. Otherwise, I'd spend most of my time accidentally crashing through walls.

SUPER-BREATH. Sounds silly, but it's surprisingly useful! It's nice to be able to put out a four-alarm fire with a single puff of breath. And when I think *freeze*, I can blow an arctic blast that covers the bad guys in ice and freezes them in their tracks.

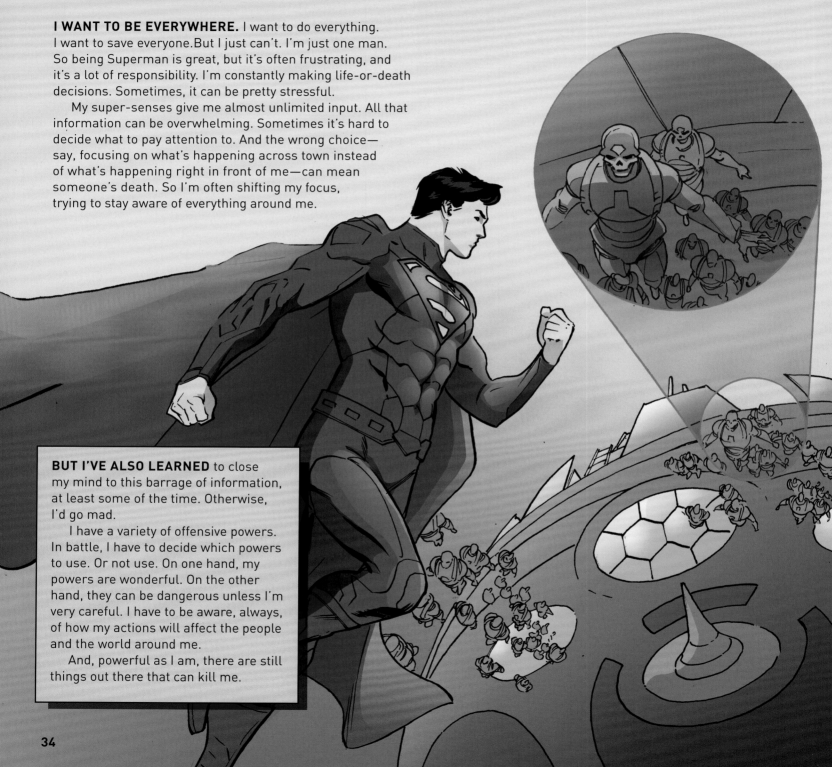

I HAVE MY LIMITS

I WANT TO BE EVERYWHERE. I want to do everything. I want to save everyone. But I just can't. I'm just one man. So being Superman is great, but it's often frustrating, and it's a lot of responsibility. I'm constantly making life-or-death decisions. Sometimes, it can be pretty stressful.

My super-senses give me almost unlimited input. All that information can be overwhelming. Sometimes it's hard to decide what to pay attention to. And the wrong choice— say, focusing on what's happening across town instead of what's happening right in front of me—can mean someone's death. So I'm often shifting my focus, trying to stay aware of everything around me.

BUT I'VE ALSO LEARNED to close my mind to this barrage of information, at least some of the time. Otherwise, I'd go mad.

I have a variety of offensive powers. In battle, I have to decide which powers to use. Or not use. On one hand, my powers are wonderful. On the other hand, they can be dangerous unless I'm very careful. I have to be aware, always, of how my actions will affect the people and the world around me.

And, powerful as I am, there are still things out there that can kill me.

MAGIC CAN AFFECT ME. Or, I guess, what we perceive as magic. Vyndktvx, who came from the fifth dimension, traveled through time, popping in and out of my life, in an attempt to destroy me.

I'm vulnerable to telepathic assault. I don't know; maybe telepathy is just another strange science-magic hybrid.

Exposure to Green Kryptonite can drastically weaken me, and given enough time—or enough Kryptonite—it would probably kill me. So far, it hasn't come to that.

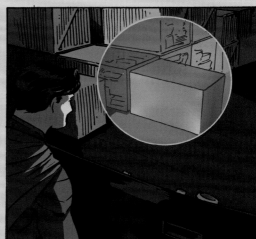

I can't see through lead. This is a limitation of sorts, but if I spot lead, it is a hint that someone might be trying to hide something from me. It just makes me want to take a closer look!

I HAVE MY OWN LITTLE PIECE OF KRYPTON HERE ON EARTH

MY FORTRESS OF SOLITUDE was created in the Arctic. It was grown from a Kryptonian crystal hidden in the ship that brought me to Earth and has helped me understand my Kryptonian ancestry. It has also been an invaluable aid in my mission to protect the people of Earth.

The Fortress is part museum. Within it are the remnants of Kryptonian civilization that were stolen from Krypton by Brainiac, including the Bottle City of Kandor.

There's also a statue of my birth mother and father, who sent me here to Earth.

In one corner, I have the portal created by my father that gives me access to the Phantom Zone.

There's even a zoo with animals from other worlds, rescued from Brainiac's ship and housed in appropriate environments.

I have a station with hundreds of screens that monitors what's happening across Earth. If I'm going to make a difference, I have to know what's going on.

I took Lois to the Fortress once to keep her safe. She said she had hoped it would be "less office-like" and more of a "vacation house" where I could relax. But part of me—my Kryptonian part, I guess—finds it restful. And, in its cold way, beautiful.

DO CLOTHES REALLY MAKE THE SUPERMAN?

MY ORIGINAL "COSTUME" wasn't really a costume at all. Just a T-shirt, jeans, and combat boots. The most important element was my cape.

When Ma and Pa pulled me from the rocket, I was swaddled in that cape. The S-shield on it is the symbol of the House of El. It belonged to my grandfather, Mon-El, and is a piece of my heritage.

It is also indestructible.

After I got to Metropolis, I traced the S-shield symbol and had Metro Copy and Print put the image on a batch of T-shirts. Inevitably, the shirts got ripped, burned, and shredded every time I went out as Superman, but the cape never even picked up a spot of soot. Kryptonian super cloth rocks!

On Brainiac's ship, I discovered a Kryptonian suit of indestructible armor. It was white when I put it on, but when it touched my skin, it analyzed and recognized my DNA and transformed, mimicking the colors of my T-shirt and jeans, and manifesting the emblem of the House of El. More Kryptonian science at work!

As Superman, I don't need indestructible battle armor. I'm nearly indestructible on my own. But it does save T-shirt bills. So, the new suit was more practical, though I had to put it on ahead of time. At first . . .

Then I discovered something awesome. As it's constructed using Kryptonian nanotechnology, it reacts not only to my skin, but also my thoughts. It contracts into the S-shield on my T-shirt when I want it to, and I can make it appear and disappear at will. Cool and convenient.

Aside from being extremely durable, it also enhances my ability to absorb solar energy and aids in my recovery from injury.

MY KRYPTONIAN FAMILY (SORT OF!)

SUPERGIRL!

HER KRYPTONIAN NAME IS Kara Zor-El, and she's my cousin.

Kara was placed in suspended animation and sent on a rocket from Krypton as a teenager when I was just a baby. Her trip through space was somehow delayed, so by the time she awakened on Earth, I was older than she was. Kara was homesick for Krypton and, at first, had trouble adjusting to being here. The Kryptonian time traveler H'el used her loneliness to try and trick her into helping him destroy the Earth.

Her powers are like mine, but she's more susceptible to sensory overload, since she hasn't had a lifetime to learn to deal with them. Still, she makes a cool ally—most of the time! Seriously, it's nice to have family around. She has her own fortress, known as Sanctuary, located in the depths of the Atlantic.

SUPERBOY!

NOT EXACTLY A RELATIVE, though a third of his DNA is Kryptonian.

Superboy is a hybrid clone created by a secret underground organization as a weapon to stop metahumans. Instead of the usual double strand of DNA, Superboy's body contains three strands: human, my Kryptonian DNA, and an unidentified third strand. Because he was a hybrid clone, Supergirl—as a pure and proud Kryptonian of the House of El—at first considered him Kon-El, "an abomination in the House of El." But her attitude changed after he fought beside us against H'el and nearly sacrificed his own life to save ours.

Superboy's powers aren't like mine—or not exactly. They're psi-based, and he's a telekinetic as well as slightly telepathic. That lets him mimic my super-strength, flight, and super-senses, and even project a kind of heat vision.

KRYPTO THE SUPERDOG!

(YOUR DOG IS A MEMBER OF YOUR FAMILY, RIGHT?)

Krypto is a Kryptonian guard hound, and he's been my family's dog for as long as I've been alive. He was pulled into the Phantom Zone when defending us from its criminals, right before I was shot into space. The beings in the Zone become ghostlike, and most are trapped there.

Krypto had followed me everywhere when I was a baby, and, even as a spirit, he found a way to follow me to Earth and protect me when he was able. A psychic told me of his existence, and I would sometimes see him at night in my dreams. I entered the Phantom Zone, found him, and released him, in his physical form, here on Earth. Krypto has Kryptonian powers, much like my own. He's smarter than an Earth dog and acts with autonomy. I'm glad to have him on my side.

EVERY WORLD HAS A FEW BAD APPLES

LIKE EARTH, KRYPTON had its criminal element. But how to deal with the worst of them? As a solution, my father invented the Phantom Zone.

The Phantom Zone is an anti-universe where Kryptonian criminals were sentenced to a total physical dematerialization and imprisoned as an alternative to death. Its prisoners became ghostly beings. Some were able to journey into the physical universe but could have no interaction with it. Ironically, the Phantom Zone survived Krypton's destruction. And, occasionally, its denizens escape . . .

GENERAL ZOD

DRU-ZOD WAS KRYPTON'S FOREMOST military leader, condemned for a failed attempt to conquer Krypton. He was even worse when he escaped to Earth, where the yellow sun gave him powers that equaled my own. He blamed my father for his incarceration and came after me, seeking revenge.

FAORA

FAORA HU-UL IS A KILLER, sentenced to the Phantom Zone for the murder of twenty-three Kryptonians. General Zod freed her from the Phantom Zone, and now she acts as his second-in-command. On Earth, she too has the advantage of superpowers derived from the rays of Earth's yellow sun.

JAX-UR

JAX-UR WAS ONCE THE FIANCÉ of my Kryptonian mother, Lara. She broke off their engagement when she learned that he'd joined a revolt against Krypton's Science Council. Jax-Ur was imprisoned in the Phantom Zone along with the other insurgents. He wants to release all the prisoners in the Phantom Zone so that they can live on Earth as gods. His dream is to convert Earth into the New Krypton.

H'EL

A CLONE CREATED FROM THE CELLS of countless Kryptonians, H'el is a sort of astronaut/time traveler who was away from Krypton when it was destroyed. He wants to go back to a time when Krypton existed and prevent its destruction. To do this, he plans to destroy Earth and use the energy generated to protect his time ship as it travels home through vast distances of time and space. In addition to Kryptonian powers, H'el possesses the abilities of teleportation, telekinesis, telepathy, and chronokinesis, the innate ability to manipulate time. With the help of Supergirl and Superboy, I was finally able to stop him.

ON EARTH, CLARK HAS PALS

I'M A STRANGER IN a land I wasn't born to, but now it's home to me. And I'm not alone. I have friends. Good friends. I need them to help me stay grounded. To really understand what I'm fighting for. I can count on their help. But if one of my enemies found out I was Superman, it could place them in a lot of danger. They're one reason I maintain my identity as Clark Kent, mild-mannered reporter.

LANA LANG is my oldest friend. We grew up together in Smallville, and she knew about my powers from the time we were kids, but she never told anyone. She's an electrical engineer now, working on jobs around the world, but we stay in touch. It's great when she passes through Metropolis.

PERRY WHITE is the editor in chief at the *Daily Planet*. He understands the requirements of modern journalism and knows how to balance the need for prof with the need for truth. I consider Perry a friend, even though he's also my boss

GEORGE TAYLOR is the *Daily Star* editor who gave me my first job as a reporter in Metropolis. The *Star* is a much smaller paper than the *Daily Planet*, but it has a lot of integrity. George encouraged me to move on to the *Planet* when the time came.

JIMMY OLSEN is a photographer for the *Daily Planet*. He's also my roommate and my best friend, and one of the few people alive who know that I'm Superman. When I hang out with Jimmy, I feel like a normal human guy.

LOIS LANE is one of my best friends and a prize-winning reporter who worked for the *Daily Planet* before its merger with Galaxy Communications. After a short stint running their TV division, she's back at the *Planet*, reporting. I knew she couldn't give up chasing the story! She's an adrenaline junkie!

RON TROUPE works for the *Planet.* He's supposed to bring a young person's perspective to his stories. So far, he's delivered. He's super-brainy and capable, and he has way more patience for Morgan Edge than I do.

NOT EVERYONE I KNOW IN METROPOLIS IS A FRIEND, OF COURSE.

GENERAL SAM LANE—now Senator Sam Lane—is Lois's father. I like Lois, so I want to like him, but there's just no way. He sees me as a danger to planet Earth and was instrumental in too many attacks on me. Lane becoming a senator might not be a good thing, but at least it gets him out of Metropolis.

CAT GRANT is a journalist who formerly wrote for the *Daily Planet.* For a while, after I quit the *Planet* and went freelance, we shared the news blog Clarkcatropolis.com. It seemed like a good idea at the time! On the surface, she seems to be more interested in style than substance, but there's a lot more to her than a sparkling personality.

MORGAN EDGE is a media mogul who owns Galaxy Communications, which acquired the *Daily Planet* a while back. Edge is in it for the money and to further his own agenda. I thought I couldn't stomach working for him, so I quit, but Perry talked me into returning to the *Planet.* Edge isn't an enemy. Yet. But he certainly isn't a friend.

LEX LUTHOR HATES MY GUTS

NEED I SAY THAT THE feeling is mutual? Alexander "Lex" Luthor is a billionaire corporatist and the world's leading scientific mind, with a PhD to prove it. At times, he's been a government consultant and military science attaché.

He's also an expert at martial arts. When I first came to Metropolis, General Lane hired Luthor to capture me. I was pinned by a runaway train—one that I finally managed to stop. It was a classic Luthor setup, in which he didn't care who else was hurt as long as he got what he wanted. And what he wanted was Supeman.

He honestly sees me as a parasitic invader masquerading as a human being. A specimen of an alien race out to destroy humanity. He actually thought a deformed calf was what my people really looked like!

Luthor is a brilliant narcissistic sociopath. Even when he's apparently doing something positive, I can't help but distrust his motives. Maybe that's because he's constantly trying to destroy me. During our time together, he created different Warsuits, some with my personal destruction clearly in mind.

This is Luthor's basic warsuit. Add a helmet and he's pretty well protected and ready for most combat situations.

He used this armor to save me from an evolved Doomsday so he could kill me himself! (Or he will use it in the future! Another time displacement experience via the Legion of Superheroes and Vyndktvx from the fifth dimension.) It's really more of a giant robot than armor. Bottom line—even in the future, Luthor will still be trying to destroy me!

He created this armor to break out of prison and "save the world." That always involves trying to kill me.

KRYPTONITE

HOW CAN EXPOSURE TO A ROCK from my home world kill me? Ask Lex Luthor.

Kryptonite crystals were originally used as a source of energy, powering Krypton's rockets, among other applications.

When the planet Krypton exploded, shards of Kryptonite flew into space. As a result, a Kryptonite crystal occasionally falls to Earth. Kryptonite wouldn't harm a native Kryptonian, but now that I've been changed by my exposure to Earth's yellow sun, I have a "very bad reaction" when I come in contact with Kryptonite.

Kryptonite comes in a rainbow of colors, or so I've been told, but the most common type is green. And even Green Kryptonite is extremely rare here. I think the biggest chunk on Earth came with me in my rocket. Just as well, since it will weaken me and possibly even kill me, if there's enough of it.

My past experience has been with Green Kryptonite, mostly when Luthor has used it against me, including the time he created the Green Kryptonite-powered K-Man. I stopped him and thought that was the end of it.

The deadliest form I've encountered is Blue Kryptonite. I was told it can "kill your spirit." Apparently that means the obliteration of someone's ghostly essence. I'm not sure what that means—can it utterly destroy a soul? Still, it makes me glad these exotic forms are almost never found on Earth.

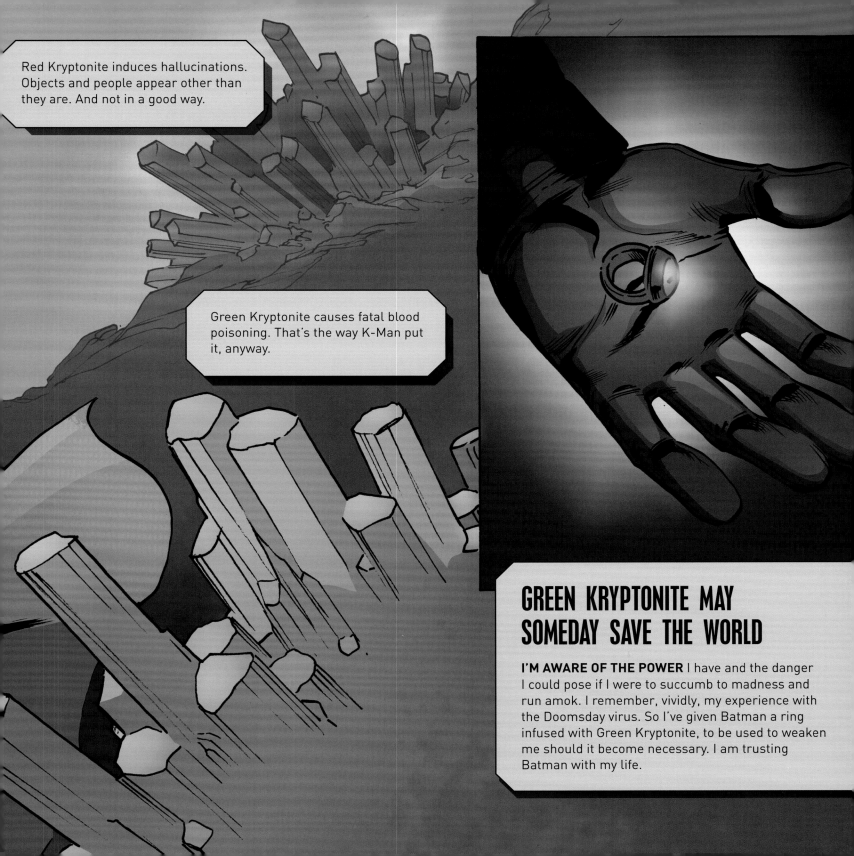

Red Kryptonite induces hallucinations. Objects and people appear other than they are. And not in a good way.

Green Kryptonite causes fatal blood poisoning. That's the way K-Man put it, anyway.

GREEN KRYPTONITE MAY SOMEDAY SAVE THE WORLD

I'M AWARE OF THE POWER I have and the danger I could pose if I were to succumb to madness and run amok. I remember, vividly, my experience with the Doomsday virus. So I've given Batman a ring infused with Green Kryptonite, to be used to weaken me should it become necessary. I am trusting Batman with my life.

LEXCORP, LUTHOR'S MONUMENT TO HIMSELF!

LEX LUTHOR HAS SPENT a lot of time and effort finding out everything he can about me—particularly the many ways he might be able to kill me. So, in between all the other things I need to do, I've made it my mission to find out all I can about him.

Like everyone, Lex Luthor has his public face, his private face, and the dark, secret place hidden in the depths of his soul. The LexCorp Building is, in some ways, analogous to Luthor himself and a monument to his own massive ego.

Some of my knowledge about LexCorp comes via the Freedom of Information Act, which gives me access to the LexCorp architectural plans. This shiny new building is a scientific and technological marvel with Luthor's penthouse office functioning as its brain-center. This is LexCorp's public face.

WHAT GOES ON BEHIND the façade is LexCorp's private face. It includes corporate offices, research labs, sections that contain proprietary research and information—things he wants to keep secret from his competitors but which aren't actually illegal.

What really interests me are Luthor's deep, dark secret spaces—his "improvements," especially, the sub-sub basement secret laboratories. Many of his more nefarious schemes are hatched in these ultra-secret places.

Several areas are lined with lead to hide specific experiments from my sight. (Lex knows what I can do. He knows I'm watching him, just as he's watching me.) I have to figure it's bad if he went to all that trouble and expense to keep me from seeing what's going on down there. It makes me wonder what he's hiding. And what horror he's going to spring on me next.

AND THEN THERE ARE MY OTHER ENEMIES

LUTHOR CAN'T TORMENT me personally all the time, so he's managed to delegate some of the responsibility. In fact, he helped create several of my major enemies. Mostly, these beings, who were once human but have now been transformed into monsters, fill me with pity. They also make me realize how easy it would be for me to step over the line and become one of them. When I'm busy, I don't think about it. But sometimes I have nightmares about what might happen if things go wrong. Of what I could become.

METALLO

Initially, John Corben was a dedicated soldier and a favorite of General Lane. Lois thinks her dad wanted her and Corben to get together, but Lois wasn't interested. When Corben realized Lois was more focused on finding Superman than being with him, he volunteered for the Steel Men project. Before this, John Corben hadn't been a bad guy. But when a super-suit was fused with his nervous system, he became Metal-Zero—aka Metallo—the anti-Superman weapon. The Metallo experience twisted Corben and, when it became clear that it would kill him, General Lane saw to it that he received a Kryptonite-powered artificial heart. That freed him from military control and made him even more of a loose cannon.

BIZARRO

Luthor created the first Bizarro by splicing my DNA with that of a teenage volunteer. Bobby is the only name I have for the boy. He was to be the first of an army of pseudo-Supermen, loyal to Luthor alone. The transformation worked, but Luthor detected weakness, so he killed the first clone and began working on a second, using more Kryptonian DNA. Creature B-Zero was everything Luthor hoped for, though some of his powers were the opposite of mine. And he wasn't too bright. But I believe Luthor developed an actual affection for Bizarro—the creature's utter devotion must have appealed to Luthor's narcissism—and he was actually saddened by Bizarro's death. I fear that, in one of LexCorp's hidden labs, a new Bizarro is now growing.

PARASITE

I can't blame the existence of the Parasite on Luthor. Joshua Michael
Allen was a bike messenger—a thoroughly unpleasant character—who
literally stumbled upon a parasitic monster. He managed to electrocute
the creature with a broken power line, but, in the process, he caught
the "Parasite" virus. He transformed into a purple monster and
became constantly hungry for life energy, which he found he could
absorb from other living beings with a touch. But the energy didn't
last and, soon, he was hungry all over again—for him an endless
kind of agony. I saved him from a suicide attempt and, when he
touched me, he absorbed my Kryptonian energy. I was weakened,
but his hunger was instantly assuaged. He grew monstrously
huge. And he manifested some of my powers. Luckily, as
he used those powers, he grew weaker, while Earth's sun
restored my own strength. In the end, I was able to capture
him. He's in Belle Reve prison now, but I know he's
constantly scheming to escape. When he does, he'll come
for me. Apparently my energy is like a drug for him.

ALIENS

HELSPONT

Helspont was a legendary conqueror from a race of beings called Daemonites. He vanquished a solar system, but his deeds inspired as much fear in his people as admiration, and so they imprisoned him. Many thousands of years later, his prison crashed to Earth. Helspont, now freed, wanted me to join him in future conquests, explaining that humanity would soon betray me as his people had betrayed him. We fought. He left. But I fear he will eventually return. Oh, and it seems he knew my father.

WRAITH

In 1938, American scientists sent a mathematical equation into space, symbolically telling whoever was out there, "Let us add up to more, together." Almost instantly, a ship crash-landed on Earth carrying the alien Wraith, sent by his people as a gift to humanity. The Army secretly used Wraith to eliminate global threats until I showed up. Wraith and I worked together, though he says he will kill me if I grow too power-hungry or pose a danger to Earth. Like me, protecting Earth is his mission. General Lane considers Wraith the "real Superman" and me the menace. In my opinion, he approves of Wraith only because he thinks he can control him.

TAKE THE FIFTH (DIMENSION)!

THE FIFTH DIMENSION EXTENDS beyond the usual three spacial dimensions to include the dimension of time. The past and the future are the same to the imps of the fifth dimension. Some of my interactions with the imps have been in my past, and I have been warned that some of them are yet to come.

VYNDKTVX

The king's official court jester, a powerful imp called Vyndktvx, loved the princess Gsptlsnz, and went (or will go!) mad with jealousy. (Dealing with the fifth dimension can get confusing.) Driven by hatred and the need to destroy everything Mxy valued, Vyndktvx used his innate ability to manipulate time to destroy my life. It took the Legion of Superheroes—humanity's descendants from the thirty-first century—to help me stop him! And it took the peculiar magic of everyone on Earth saying their *own* names backwards to send Vyndktvx back to the fifth dimension.

MRS. NYXLY

The first imp I met was Mrs. Nyxly. She was my landlady when I first arrived in Metropolis and, even at the time, I realized there was something uncanny about her. It turned out that in the fifth dimension, she was the princess Gsptlsnz, and she had come to Earth specifically to help protect me. But protect me from what? It's a long story.

MR. MXYZPTLK

Mxy is (or will be) a trickster—practical joker who won (or will win) the admiration of his king by making him laugh. And the things the king will laugh hardest at, apparently, are the tricks Mxy will play on me. (You can't imagine how I'm looking forward to all of this!) For this alone, Mxy valued (or will value) me. Princess Gsptlsnz fell in love with Mxy. And married him. And he followed her to Earth, where they had a loving human marriage until he died. Although . . . can an imp really die?

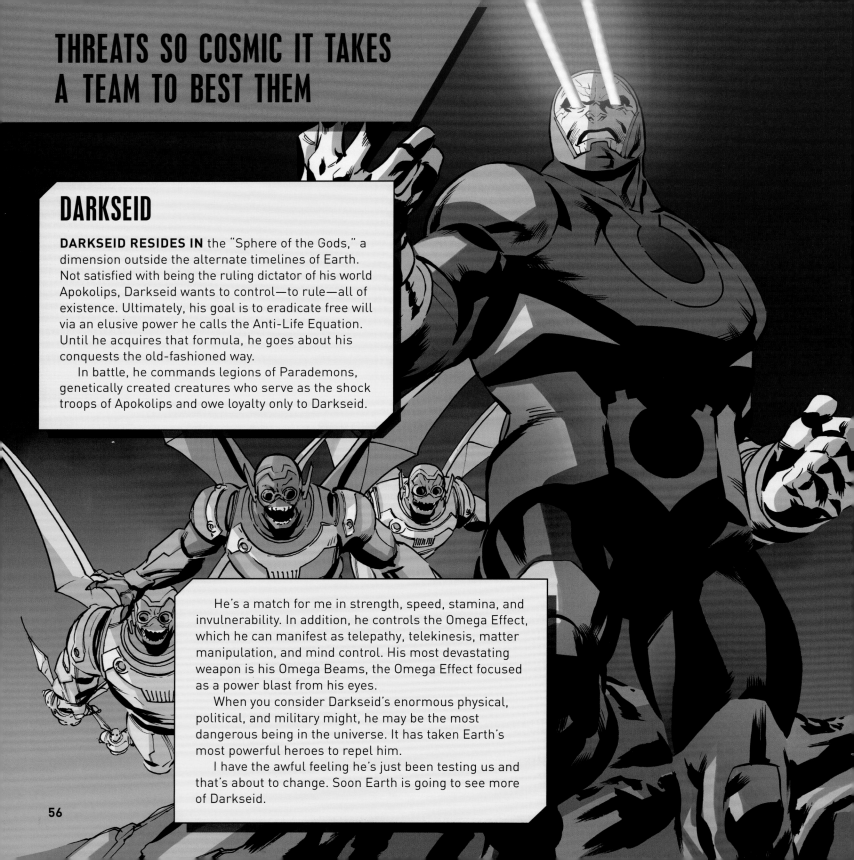

DARKSEID

DARKSEID RESIDES IN the "Sphere of the Gods," a dimension outside the alternate timelines of Earth. Not satisfied with being the ruling dictator of his world Apokolips, Darkseid wants to control—to rule—all of existence. Ultimately, his goal is to eradicate free will via an elusive power he calls the Anti-Life Equation. Until he acquires that formula, he goes about his conquests the old-fashioned way.

In battle, he commands legions of Parademons, genetically created creatures who serve as the shock troops of Apokolips and owe loyalty only to Darkseid.

He's a match for me in strength, speed, stamina, and invulnerability. In addition, he controls the Omega Effect, which he can manifest as telepathy, telekinesis, matter manipulation, and mind control. His most devastating weapon is his Omega Beams, the Omega Effect focused as a power blast from his eyes.

When you consider Darkseid's enormous physical, political, and military might, he may be the most dangerous being in the universe. It has taken Earth's most powerful heroes to repel him.

I have the awful feeling he's just been testing us and that's about to change. Soon Earth is going to see more of Darkseid.

DOOMSDAY

IN SOME WAYS, DOOMSDAY is my worst nightmare.

Essentially, it is a genetically created, naturally regenerative corporate monster from a parallel reality, created to attack and destroy the competition. (Of course, Vyndktvx had a hand in creating it!) It attacked Krypton eons ago and was defeated and imprisoned in the Phantom Zone. But as the walls of the Phantom Zone began to weaken, Doomsday escaped to wreak havoc on the Earth. Doomsday is super-strong and invulnerable—actually able to withstand Darkseid's Omega Beams—and determined to annihilate life on Earth.

When I fought Doomsday, it generated a corrosive field that destroyed everything around it. Only I could directly confront Doomsday. With the aid of Steel and the Justice League, I managed to rip it in half and incinerate it. To prevent its regeneration, I inhaled its ashes. Bad idea. I began to turn into Superdoom. I became my worst nightmare.

Finally, the Justice League devised a way to save me, but it was a close call. Somewhere in the back of my mind, I'm afraid Doomsday will return. Part of me fears that monster will always be a part of me. This experience is the kind of thing I always feared might happen. In my nightmares, I have seen the enemy—and it is me.

HERE'S WHERE THE JUSTICE LEAGUE COMES IN

AS SUPERMAN, I HAVE ALLIES— the Justice League. We banded together to protect the Earth, and we have each other's backs.

The Watchtower, Justice League headquarters, is a satellite orbiting 22,300 miles above the Earth. From there, we monitor Earth and, working together, we've dealt with some of the greatest threats to humanity that have ever existed, including Doomsday and Darkseid.

GREEN LANTERN

He has the power in that ring, and he can control it! The will is in the man.

BATMAN

I know I think fast, but Batman thinks more strategically. He keeps a very cool head when a cool head is needed. I count on him and trust him implicitly.

CYBORG

I appreciate his way with tech! I'd love to see him deal with Kryptonian tech. I'll have to get him to the Fortress soon.

AQUAMAN

It's good to have an underwater specialist, especially since the Earth is two-thirds water! But he carries his weight on land and in the air.

THE FLASH

Which of us is faster? Always open to debate. But in a flat-out race, I know I'd win. Of course, he might disagree . . .

WONDER WOMAN

Power! Brains! Passion! Will! Strength! It's an unbeatable combination, as many a bad guy has found out!

Okay, these remarks about my teammates are kind of flippant. But it's hard to say what's in my heart. These people have been with me through some of the most difficult times of my life. And together, we've saved Earth several times over.

The bottom line is, I'm always worried that some evil power will overwhelm me and turn my strength against the world I love. But with the Justice League around me, I know I can relax. If the worst happens, they'll do their best to save me— and to stop me if they have to. What more can you ask of friends?

STEEL

Sometimes even the Justice League can't do it all. That's when other heroes step in. Steel has been one of them. When I first met him, he was working for Luthor, but he came to his senses when he saw the kind of man Luthor really is. Since then, he forged his own armor and aided me against Metallo and Doomsday, among others.

WONDER WOMAN: MORE THAN AN ALLY...

EVEN IN MY CLARK PERSONA, I'm careful when dealing with people. I have to be. Compared to me, humans are fragile. Any one of my powers could hurt someone, or even kill them. Heck, even an association with me can put them in harm's way.

So it's wonderful to be around Wonder Woman. Sure she's proud. Sure she thinks she's always right. But she's also loyal. And protective. And determined to do the right thing. She won't let me be anything but my best self.

I don't have to be careful around her. I don't have to hold back. Physically, she's as tough as I am. She's a fierce, amazing fighter, a true Amazon in every sense of the word. And her fierceness brings out a matching fierceness in me.

I love being with her in battle. I love talking to her. Heck, I even love arguing with her. She and I are both something new on Earth. It's great to be able to share the experience with someone who understands.

LOIS IS A GREAT REPORTER. Crusading. Tenacious. Brilliant. Usually ten steps ahead of the rest of us. Like me, she wants to do what's right. She's utterly ethical and fearless in following a lead. She wants her stories to mean something. She wants them to help people. She wants the truth. On the other hand, she has been known to bend the rules when chasing a story. Which has, occasionally, put her in danger. Not that she lets that stop her. It's just that her common sense sometimes takes a backseat to the thrill of the chase. We butt heads. We argue. She always thinks she's right. Hmmm . . . I'm seeing a pattern here. I guess I like fierce women.

She's my friend. Sometimes I wish she were more. But she has a boyfriend. I'm not sure of her feelings for me—as Clark or as Superman. And she's so perceptive. I sometimes wonder if Lois secretly suspects that I lead a double life. I've told Jimmy about my secret identity. I rationalized it by telling myself he's my best friend and my roommate. It would just be too hard to keep it a secret. But I haven't told Lois. Why? Time and again, she's been endangered, just because she's written stories that focus on Superman. And maybe with Lois, I like the time I spend with her as Clark and the feeling that, at heart, I'm human.

SO WHO AM I?

I'M KRYPTONIAN. I'M HUMAN. I'm a citizen of Earth, my adopted world. And a member of a larger universe made up of myriad worlds and dimensions.

Ma and Pa—down-to-earth Kansas farmers—took me in when I was orphaned and alone. They protected me. And when it was evident that I had been given power beyond anything they could have imagined, Ma and Pa taught me to use that power to help people.

So now I choose to be Earth's protector. And if sometimes, my very existence is a danger to the people around me, I'll do my best to keep humanity safe. To make the positive things I bring to Earth far outweigh the negative. To try to be a force for good in an increasingly dangerous world.

I realize it's not just me. There seem to be ever more super-powered beings on Earth—some good, some evil. It's up to me to be one of the good guys. To keep Metropolis and its citizens safe. To keep my adopted planet safe. To protect the yellow sun that Earth orbits! Heck, to try to keep the universe safe, too.

It's a big job. I'll do my best to honor it as Clark Kent— and as Superman.

INSIGHT
EDITIONS

PO Box 3088
San Rafael, CA 94912
www.insighteditions.com

 Find us on Facebook: www.facebook.com/InsightEditions

 Follow us on Twitter: @insighteditions

Library of Congress Cataloging-in-Publication Data available.

ISBN: 978-1-60887-491-0

Publisher: **Raoul Goff**
Acquisitions Manager: **Robbie Schmidt**
Art Director: **Chrissy Kwasnik**
Designer: **Ashley Quackenbush**
Executive Editor: **Vanessa Lopez**
Project Editor: **Chris Prince**
Production Editor: **Elaine Ou**
Editorial Assistant: **Katie DeSandro**
Production Manager: **Blake Mitchum**

Illustrations: **Marcus To**
Colors: **Greg Menzie & Irma Kniivila**

INSIGHT EDITIONS would like to thank Josh Anderson, Elaine Piechowski,
Patrick Flaherty, Kevin Kiniry, Freddie E. Williams II, and Leah Bloise.

ROOTS of PEACE   ❀ REPLANTED PAPER

Insight Editions, in association with Roots of Peace, will plant two trees for each tree used in
the manufacturing of this book. Roots of Peace is an internationally renowned humanitarian
organization dedicated to eradicating land mines worldwide and converting war-torn lands
into productive farms and wildlife habitats. Roots of Peace will plant two million fruit and
nut trees in Afghanistan and provide farmers there with the skills and support necessary for
sustainable land use.

Manufactured in China by Insight Editions

10 9 8 7 6 5 4 3 2 1